Outgrowing God

By Richard Dawkins

Outgrowing God
Science in the Soul
Brief Candle in the Dark
An Appetite for Wonder
The Magic of Reality *(with Dave McKean)*
The Greatest Show on Earth
The God Delusion
The Ancestor's Tale
A Devil's Chaplain
Unweaving the Rainbow
Climbing Mount Improbable
River Out of Eden
The Blind Watchmaker
The Extended Phenotype
The Selfish Gene

RICHARD DAWKINS

Outgrowing

God

A BEGINNER'S GUIDE

RANDOM HOUSE

NEW YORK

Published in the United States by Random House, an imprint and
division of Penguin Random House LLC, New York.

RANDOM HOUSE and the HOUSE colophon are registered
trademarks of Penguin Random House LLC.

Originally published in hardcover in Great Britain by Bantam
Press, an imprint of Transworld Publishers, a Penguin
Random House UK Company, in 2019.

Photo credits are located on page 279.

LIBRARY OF CONGRESS CATALOGING-IN-PUBLICATION DATA
Names: Dawkins, Richard, author.
Title: Outgrowing God : a beginner's guide / Richard Dawkins.
Description: New York : Random House, 2019. | Includes index.
Identifiers: LCCN 2019016019 | ISBN 9781984853912 |
ISBN 9781984853936 (ebook)
Subjects: LCSH: Atheism.
Classification: LCC BL2747.3 .D385 2019 | DDC 211/.8—dc23
LC record available at https://lccn.loc.gov/2019016019

Printed in the United States of America on acid-free paper

randomhousebooks.com

2 4 6 8 9 7 5 3 1

First U. S. Edition

Chapter illustrations by Jana Lenzova
Line illustrations by Global Blended Learning Ltd.

For William
And all young people when they're old enough to decide
for themselves

Contents

PART ONE

Goodbye God

· 1 ·

So many gods!

Do you believe in God?

Which god?

Thousands of gods have been worshipped throughout the world, throughout history. Polytheists believe in lots of gods all at the same time (*theos* is Greek for 'god' and *poly* is Greek for 'many'). Wotan (or Odin) was the chief god of the Vikings. Other Viking gods were Baldr (god of beauty), Thor (the thunder god with his mighty hammer) and his daughter Throd. There were goddesses like Snotra (goddess of wisdom), Frigg (goddess of motherhood) and Ran (goddess of the sea).

The ancient Greeks and Romans were also polytheistic. Their gods, like the Viking ones, were very human-like, with powerful human lusts and emotions. The twelve Greek gods and goddesses are often paired with Roman equivalents who were thought to do the same jobs, such as Zeus (Roman Jupiter), king of the gods, with his thunderbolts; Hera, his wife (Juno); Poseidon (Neptune), god of the sea; Aphrodite (Venus), goddess of love; Hermes (Mercury), messenger of the gods, who flew on winged sandals; Dionysos (Bacchus), god of wine. Of the major religions that survive today, Hinduism is also polytheistic, with thousands of gods.

Countless Greeks and Romans thought their gods were real – prayed to them, sacrificed animals to them, thanked them for good fortune and blamed them when things went wrong. How do we know those ancient people weren't right? Why does nobody believe in Zeus any more? We can't know for sure, but most of us are confident enough to say we are 'atheists' with respect to those old gods (a 'theist' is somebody who believes in god(s) and an 'atheist' – a-theist, the 'a' meaning 'not' – is someone who doesn't). Romans at one time said the early Christians were atheists because they didn't believe in Jupiter or Neptune or any of that crowd. Nowadays we use the word for people who don't believe in any gods at all.

Like you I expect, I don't believe in Jupiter or Poseidon or Thor or Venus or Cupid or Snotra or Mars or Odin or Apollo. I don't believe in ancient Egyptian gods like Osiris, Thoth, Nut, Anubis or Horus his brother who, like Jesus and many other gods from around the world, was said to have been born to a virgin. I don't believe in Hadad or Enlil or Anu or Dagon or Marduk or any of the ancient Babylonian gods.

I don't believe in Anyanwu, Mawu, Ngai or any of the sun gods of Africa. Nor do I believe in Bila, Gnowee, Wala, Wuriupranili or Karraur or any of the sun goddesses of Australian aboriginal tribes. I don't believe in any of the many Celtic gods and goddesses, such as Edain the Irish sun goddess or Elatha the moon god. I don't believe in

Mazu the Chinese water goddess or Dakuwaqa the Fijian shark god, or Illuyanka the Hittite dragon of the ocean. I don't believe in any of the hundreds and hundreds of sky gods, river gods, sea gods, sun gods, star gods, moon gods, weather gods, fire gods, forest gods ... so many gods to not believe in.

And I don't believe in Yahweh, the god of the Jews. But it's quite likely you do, if you were brought up a Jew, a Christian or a Muslim. The Jewish god was adopted by the Christians and (under the Arabic name, Allah) the Muslims. Christianity and Islam are offshoots of the ancient Jewish religion. The first part of the Christian Bible is purely Jewish, and the Muslim holy book, the Quran, is partly derived from Jewish scriptures. Those three religions, Judaism, Christianity and Islam, are often grouped together as the 'Abrahamic' religions, because all three trace back to the mythical patriarch Abraham, who is also revered as the founder of the Jewish people. We'll meet Abraham again in a later chapter.

All those three religions are called monotheistic because their members claim to believe in only one god. I say 'claim to' for various reasons. Yahweh, today's dominant god (whom I'll therefore spell with a capital G, God) started out in a small way as the tribal god of the ancient Israelites who, they believed, looked after them as his 'chosen people'. (It's a historical accident – the adoption of Christianity as the Roman Empire's official religion by the Emperor Constantine in AD 312 – that led to Yahweh's being worshipped around the world today.) Neighbouring tribes had their own gods who, they believed, gave *them* special protection. And although the Israelites worshipped their own tribal god Yahweh, they didn't necessarily disbelieve in the gods of rival tribes, such as Baal, the fertility god of the Canaanites; they just thought Yahweh was more powerful – and also extremely jealous (as we shall see later on): woe betide you if he caught you flirting with any of the other gods.

The monotheism of modern Christians and Muslims is also rather dubious. For example, they believe in an evil 'devil' called Satan (Christianity) or Shaytan (Islam). He goes under a variety of other names too, such as Beelzebub, Old Nick, the Evil One, the Adversary, Belial, Lucifer. They wouldn't call him a god, but they regard him as having god-like powers and he is seen, with his forces of evil, as waging a titanic war against the good forces of God. Religions often inherit ideas from older religions. The notion of a cosmic war of good versus evil probably comes from Zoroastrianism, an early religion founded by the Persian prophet Zoroaster, which influenced the Abrahamic religions. Zoroastrianism was a two-gods religion, the good god (Ahura Mazda) battling it out with the evil god (Angra Mainyu). There are still a few Zoroastrians about, especially in India. That's yet another religion I don't believe in and probably you don't either.

One of the weirder accusations levelled at atheists, especially in America and Islamic countries, is that they worship Satan. Of course, atheists don't believe in evil gods any more than they believe in good ones. They don't believe in anything supernatural. Only religious people believe in Satan.

Christianity verges on polytheism in other ways, too. 'Father, Son and Holy Spirit' are described as 'three in one and one in three'. Exactly what this means has been disputed, often violently, down the centuries. It sounds like a formula for squeezing polytheism into monotheism. You could be forgiven for calling it tri-theism. The early split in Christian history between the Eastern (Orthodox) and Western (Roman) Catholic Church was largely caused by a dispute over the following question: Does the Holy Ghost 'proceed from' (whatever that might mean) the Father *and* the Son, or just from the Father? That really is the kind of thing theologians spend their time thinking about.

And then there's Jesus's mother, Mary. For Roman Catholics, Mary is a goddess in all but name. They deny that she is a goddess, but they still pray to her. They believe she was 'immaculately conceived'. What does that mean? Well, Catholics believe we are all 'born in sin'. Even tiny babies who, you might think, are a bit young to sin. Anyway, Catholics think Mary (like Jesus) was an exception. All the rest of us inherit the sin of Adam, the first man. In fact, Adam never actually existed, so he couldn't sin. But Catholic theologians aren't put off by little details like that. Catholics also believe that Mary, instead of dying like the rest of us, was sucked bodily 'up' into heaven. They portray her as the 'Queen of Heaven' (sometimes even 'Queen of the Universe'!) with a little crown balanced on top of her head. All those things would seem to make her at least as much of a goddess as any of the thousands and thousands of Hindu deities (which Hindus themselves say are just different versions of one single god). If the Greeks, Romans and Vikings were polytheistic, then Roman Catholics are too.

Roman Catholics also pray to individual saints: dead people who are regarded as especially holy, and have been 'canonized' by a Pope. Pope John Paul II canonized 483 new saints, and Francis, the current pope, canonized no fewer than 813 on one day alone. Many of the saints are thought to have special skills, which make them worth praying to for particular purposes or particular groups of people. Saint Andrew is the patron saint of fishmongers, Saint Bernward the patron saint of architects, Saint Drogo the patron saint of coffee-house owners, Saint Gummarus the patron saint of lumberjacks, Saint Lidwina the patron saint of ice-skaters. If you need to pray for patience, a Catholic might advise you to pray to Saint Rita Cascia. If your faith is shaky, try Saint John of the Cross. If in distress or mental anguish, Saint Dymphna might be your best bet. Cancer sufferers tend to try Saint Peregrine. If you've lost your keys, Saint Anthony is your man. Then there are the angels, who come in various ranks, from seraphs at the top, down through archangels to your personal guardian angel. Again, Roman Catholics will deny that angels are gods or demigods, and they will protest that they don't really pray to saints but just ask them to put in a good word with God. Muslims, too, believe in angels. Also in demons, which they call djinns.

I don't think it matters much whether Mary and the saints and archangels and angels are gods or demigods or neither. Arguing over whether angels are demigods is rather like arguing whether fairies are the same as pixies.

Although you presumably don't believe in fairies and pixies, it is quite likely you have been brought up in one of the three Abrahamic faiths as Jewish, Christian or Muslim. As it happens I was brought up Christian. I went to Christian schools and was confirmed in the Church of England aged 13. I finally gave up Christianity when I was 15. One of the reasons I gave it up was this. I had already worked out when I was about nine that if I'd been born to Viking parents I'd firmly believe in Odin and Thor. If I'd been born in ancient Greece I'd worship Zeus and Aphrodite. In modern times, if I'd been born in Pakistan or Egypt I'd believe that Jesus was only a prophet, not the Son of God as the Christian priests teach. If I'd been born to Jewish parents I'd still be waiting for the Messiah, the long-promised saviour, instead of believing that Jesus was the Messiah as my Christian schools taught. People growing up in different countries copy their parents and believe in the god or gods of their own country. These beliefs contradict each other, so they can't all be right.

If one of them is right, why should it be the belief that you happen to have inherited in the country where you were born? You don't have to be very sarcastic to think something like this: 'Isn't it remarkable that almost every child follows the same religion as their parents, and it always just happens to be the right religion!' One of my pet peeves is the habit of labelling young children with the religion of their parents: 'Catholic child', 'Protestant child', 'Muslim child'. Such phrases can be heard used of children too young to talk, let alone to hold religious opinions. It seems to me as absurd as talking about a 'Socialist child' or 'Conservative child', and nobody would ever use a phrase like that. I don't think we should talk about 'atheist children' either.

Now, a few more names for people who don't believe. There are many who prefer to avoid the word 'atheist', even though they don't believe in any named gods. Some simply say 'I don't know, we can't know.' These people often call themselves 'agnostics'. The word (based on a Greek word meaning 'unknowing') was coined by Thomas Henry Huxley, a friend of Charles Darwin known as 'Darwin's Bulldog' because he fought for Darwin in public when Darwin was too shy, too busy or too ill to do so. Some people who call themselves agnostic think it's equally likely that gods do, or do not, exist. I think that's rather feeble, and Huxley would have agreed. We can't prove there are no fairies but that doesn't mean we think there's a 50:50 chance fairies exist. More sensible agnostics say they don't know for sure, but think it's pretty unlikely any sort of god exists. Other agnostics might say it's not unlikely but we just don't know.

There are people who don't believe in named gods but still hanker after 'some sort of higher power', a 'pure spirit', a creative intelligence about which we know nothing except that it designed the universe. They might say something like: 'Well, I don't believe in God' – probably meaning the Abrahamic God – 'but I can't believe this is all there is. There must be something more, something beyond.'

Some of these people call themselves 'pantheists'. Pantheists are a little vague about what they believe. They say things like 'My god is everything' or 'My god is nature' or 'My god is the universe'. Or 'My god is the deep mystery of everything we don't understand'. The great Albert Einstein used the word 'God' in pretty much this last sense. That's very different from a god who listens to your prayers, reads your innermost thoughts and forgives (or punishes) your sins – all of which the Abrahamic God is supposed to do. Einstein was adamant that he didn't believe in a personal god who does any of those things.

Others call themselves 'deists'. Deists don't believe in any of the thousands of named gods of history. But they believe in something a little more definite than pantheists do. They believe in a creative intelligence who invented the laws of the universe, set everything in motion at the beginning of time and space, and then sat back and did nothing more: just let everything happen according to the laws that he (it?) had laid down. Several of the founding fathers of the United States, men like Thomas Jefferson and James Madison, were deists. My suspicion is that, if they'd lived after Charles Darwin instead of in the eighteenth century, they'd have been atheists, but I can't prove it.

When people say they are atheists they don't mean they can prove that there are no gods. Strictly speaking, it's impossible to prove that something does *not* exist. We don't positively know there are no gods, just as we can't prove there are no fairies or pixies or elves or hobgoblins or leprechauns or pink unicorns; just as we can't prove that Santa Claus or the Easter Bunny or the Tooth Fairy don't exist. There's a billion things you can imagine and nobody can disprove. The philosopher Bertrand Russell made the point with a vivid word picture. If I were to tell you, he said, that there is a china teapot in orbit around the sun, you could not disprove my claim. But failure to disprove something is not a good reason to believe it. In some strict sense we should all be 'teapot agnostics'. In practice we are a-teapotists. You can be an atheist in the same (technically agnostic) way you're an a-teapotist, an a-fairyist, an a-pixieist, an a-unicornist, an a-anything-you-might-dream-up-ist.

Strictly speaking, we should all be agnostic about all those billions of things we can imagine and nobody can disprove. But we don't *believe* in them. And until somebody offers a reason to believe, we are wasting our time bothering to do so. That is the approach we all take to Thor and Apollo and Ra and Marduk and Mithras and the Great Juju up the Mountain. Couldn't we go a little further and think the same way about Yahweh or Allah?

'Until somebody offers a reason,' I said. Well, plenty of people have offered what they thought were reasons for believing in one god or another. Or for believing in some kind of un-named 'higher power' or 'creative intelligence'. So we need to look at those reasons and see whether they really are good reasons. We'll see some of them in the course of this book. Especially in Part Two, which discusses evolution.

On that huge subject, all I need say at present is that evolution is a definite fact: we are cousins of chimpanzees, slightly more distant cousins of monkeys, very much more distant cousins of fish and so on.

Many people believe in their god or gods because of scripture: the Bible, the Quran or some other holy book. This chapter might already have prepared you to doubt that reason for belief. There are so many different faiths. How do you know the holy book you have been brought up with is the true one? And if all the others are wrong, what makes you think your holy book isn't wrong too? Many of you reading this may have been brought up on one particular holy book, the Bible of the Christians. The next chapter will be about the Bible. Who wrote it, and what reason has anyone to believe that what it says is true?

· 2 ·

But is it true?

How much of what we read in the Bible is true?

How do we know anything in history really happened? How do we know Julius Caesar existed? Or William the Conqueror? No eye-witnesses survive; and even eye-witnesses can be surprisingly unreliable, as any police officer collecting statements will tell you. We know that Caesar and William existed, because archaeologists have found tell-tale relics and because there's lots of confirmation from documents written when they were alive. But when the only evidence for an event or person wasn't written down until decades or centuries after the death of any witnesses, historians get suspicious. The evidence is weak because it was passed on by word of mouth and could easily become distorted. Especially if the writer was biased. Winston Churchill said: 'History will be kind to me. I intend to write it!' We'll see in this chapter that there are problems with most of the stories about Jesus in the New Testament. The Old Testament must wait till Chapter 3.

Jesus would have spoken Aramaic, a Semitic language related to Hebrew. The books of the New Testament were originally written in Greek; those of the Old Testament in Hebrew. Many English translations exist.

The most famous is the King James version of 1611, so called because it was commissioned by King James I of England (James VI of Scotland). The King James version is the translation I prefer because the language is beautiful – not surprisingly, since its English is the English of Shakespeare's time. However, because that language is not always so clear to modern readers, in this book I have reluctantly decided to use a modern translation, the New International version; quotations are from the latter unless otherwise stated.

There's a party game called Chinese Whispers (in Britain) or Telephone (in America). You line up, say, ten people in a row. The first person whispers something – it might be a story – to the second. The second whispers the story to the third, the third person to the fourth, and so on. Finally, when the story reaches the tenth person she repeats what she has heard to the whole party. Unless the original story was exceptionally simple and brief, it will have become greatly changed, often in a funny way. It's not just the words that change down the line, but important details of the story itself.

Before writing was invented and before scientific archaeology started, word-of-mouth storytelling, with all its Chinese Whispery distortions, was the only way people learned about history. And it's terribly unreliable. As each generation of storytellers gives way to the next, the story becomes more and more garbled. Eventually, history – what actually happened – becomes lost in myth and legend. It's difficult to know whether there ever was a real person behind the legendary Greek hero Achilles, or the fabled beauty Helen whose face 'launched a thousand ships'. When the poet Homer finally wrote the stories down (and we don't know when that was, even to the nearest century) they'd been distorted through generations of word-of-mouth retelling. Any reliable truth had dissolved away. We don't know who 'Homer' was or when he lived; whether he was blind, as legend has it; whether he was one person or many. And we don't know how his stories originally began, before they passed through the distorting filter of word-of-mouth retelling. Did they start as factual accounts and then become garbled? Or did they start as made-up fiction and get changed in the retelling?

The same applies to the stories in the Old Testament. We have no more reason to believe them than we do Homer's stories about Achilles or Helen. The stories of Abraham and Joseph are Hebrew legends, just as Homer's are Greek legends. How about the New Testament? It offers a better hope of finding true history because it refers to a more recent period than the Old: a mere two thousand years ago. But how much do we really know about Jesus? Can we be sure he even existed? Most, though not all, modern scholars think he probably did. What evidence do we have?

The gospels? They're printed at the beginning of the New Testament, so you might think they were written first. Actually, the oldest books in the New Testament come near the end: the letters of St Paul. Unfortunately, Paul says hardly anything about Jesus's life. There's lots about the religious meaning of Jesus, especially his death and resurrection. But almost nothing that even claims to be history. Maybe Paul thought his readers already knew the story of Jesus's life. But it's possible Paul didn't know it himself: remember, the gospels were not yet written. Or maybe he didn't think it was even important. This lack of facts about Jesus in Paul's letters makes historians wonder. Isn't it a little odd that Paul, who wanted people to worship Jesus, says almost nothing about what Jesus actually said or did?

Another thing that worries historians is that there are hardly any mentions of Jesus in histories outside the gospels. The Jewish historian Josephus (AD 37–c.100), writing in Greek, had only this to say:

> About this time there lived Jesus, a wise man, if indeed one ought to call him a man. For he was one who performed surprising deeds and was a teacher of such people as accept the truth gladly. He won over many Jews and many of the Greeks. He was the Messiah. And when, upon the accusation of the principal men among us, Pilate had condemned him to a cross, those who had first come to love him did not cease. He appeared to them spending a third day restored to life, for the prophets of God had foretold these things and a thousand other marvels about him. And the tribe of the Christians, so called after him, has still to this day not disappeared.

Many historians suspect this passage is a forgery, stuck in later by a Christian writer. The most suspicious phrase is 'He was the Messiah.' In the Jewish tradition, 'Messiah' was the name given to the long-promised Jewish king or military leader who would be born to triumph over the enemies of the Jews. Christians taught that Jesus was the Messiah ('Christ' is simply the Greek translation of this word). But to a devout Jew, Jesus didn't look at all like a military leader. In fact, that's putting it mildly. His message of peace – even turning the other cheek when somebody hits you – is not what we expect of a soldier. And far from leading the Jews against the Roman oppressors of his time, Jesus went meekly to his execution at their hands. The idea that Jesus was the Messiah would have seemed pretty bonkers to a devout Jew like Josephus. If Josephus had somehow gone against his whole upbringing and convinced himself that so unlikely a character as Jesus was the Messiah, he would have made a big song and dance of it. He wouldn't have just dropped in a casual 'He was the Messiah'. It does sound very like a later Christian forgery. That's certainly what most scholars now believe.

The only other early historian who mentions Jesus is the Roman Tacitus (AD 54–120). His writing offers more convincing evidence for Jesus's existence, for the back-handed reason that Tacitus has nothing good to say about Christians. Writing in Latin about an event during the persecution of the early Christians by the Emperor Nero (AD 37–87), Tacitus said:

> Nero fastened the guilt and inflicted the most exquisite tortures on a class hated for their abominations, called Christians by the populace. Christus, from whom the name had its origin, suffered the extreme penalty during the reign of Tiberius at the hands of one of our procurators, Pontius Pilatus, and a most mischievous superstition, thus checked for the moment, again broke out not only in Judæa, the first source of the evil, but even in Rome, where all things hideous and shameful from every part of the world find their centre and become popular.

No later insertion by Christians here!

The balance of probability, according to most but not all scholars, suggests that Jesus did exist. Of course we'd know for sure, if we could be certain the four gospels of the New Testament were historically true. Until recently, nobody doubted them. There's even a proverbial phrase in English, 'gospel truth', meaning as true as true can be. But that phrase rings rather hollow today, after studies in the nineteenth and twentieth centuries by (especially German) scholars.

Who wrote the gospels? And when? Many people wrongly believe that the gospel of 'Matthew' was written by Matthew the tax-collector, one of Jesus's twelve close companions. And that the gospel of 'John' was written by another of that small group, the John who came to be known as 'the beloved disciple'. They think 'Mark' was written by a young companion of Jesus's chief disciple Peter, and 'Luke' by a doctor friend of Paul. But nobody has the faintest idea who really wrote the gospels. We have no convincing evidence in any of the four cases. Later Christians simply stuck a name on the top of each gospel for convenience. It must have seemed better than giving them dull, neutral labels like A, B, C and D. No serious scholar today thinks the gospels were written by eye-witnesses, and all agree that even Mark, the oldest of the four gospels, was written about 35 or 40 years after the death of Jesus. Luke and Matthew derived most of their stories from Mark, plus some from a lost Greek document known as 'Q'. Everything that is in the gospels suffered from decades of word-of-mouth retelling, Chinese-Whispery distortion and exaggeration before those four texts were finally written down.

The assassination of President Kennedy in 1963 was witnessed by hundreds of people. It's on film. Newspapers around the world reported it the very same day. A committee called the Warren Commission was set up to examine every detail of what happened. It took expert advice from scientists, doctors, forensic detectives and fire-arms experts. The 888-page Warren Report concluded that Kennedy was shot by Lee Harvey Oswald, acting alone. But over the years myths and legends and conspiracy theories have arisen, and they'll probably grow with the telling long after all the eye-witnesses are dead.

The '9/11' attacks on New York and Washington DC took place less than 20 years ago, a shorter time than elapsed between the death of Jesus and the writing of the oldest gospel, Mark. The facts of 9/11 have been massively documented, reported by multiple witnesses and chewed over in minute detail ever since. Yet they are not agreed. The internet is abuzz with contradictory rumours, legends and theories. Some people think it was an American plot. Or an Israeli plot. Even a plot by aliens from outer space. Others at the time thought, with no evidence, that it was masterminded by Saddam Hussein, dictator of Iraq. This justified, in their eyes, President Bush's invasion of that country (though that was never the official reason). Eye-witnesses photographed what they thought was the face of Satan in the smoky dust clouds hanging over New York that day.

It's unfortunately true – and the internet brings it home as never before – that people simply make stuff up. And rumours and gossip spread like epidemics, regardless of truth. The great American author Mark Twain is supposed to have said: 'A lie can spread half way around the world while the truth is putting on its shoes.' And not only malicious lies, but good stories that aren't true but are amusing and fun to recount, especially if you were told them in good faith and don't positively know they're untrue. Or stories that, if not amusing, are spookily uncanny – another reason why so many are passed on.

Here's a typical example of how an untrue story spreads because it's entertaining and fits with people's expectations or prejudices. First some background. You may have heard of 'the Rapture'. Some preachers and writers, drawing on particular passages in the Bible, have recently revved up thousands of people, mostly in America, to believe that soon a few lucky ones, chosen for their goodness, will suddenly shoot up into the sky and disappear into heaven. This 'Rapture' will herald the promised 'Second Coming' of Jesus. The rest of us – unraptured – will be 'Left Behind'. People we know will suddenly vanish without trace. Presumably 'up into the sky' means that raptured Australians will shoot in the opposite direction from raptured Europeans!

Now here's the story I mentioned. It isn't true but it's widely believed, and it shows how a good story will spread. A woman from Arkansas was driving behind a truck carrying a load of life-sized human-shaped balloons. The truck crashed and the pink inflated dolls floated skywards because they'd been inflated with helium. Thinking she was witnessing the Rapture and the Second Coming of Jesus, the woman screamed, 'He's back, he's back!' and climbed out through the sunroof of her moving car, in order to be raptured up to heaven. The resulting 20-car pile-up killed 13 innocent people as well as the woman herself. Note the spurious precision of that '13 innocent people'. You might think that a mere rumour wouldn't specify such detail. But you'd be wrong.

And you can see how 'spreadable' that story is. If somebody told it to you as fact, you'd almost certainly rush to tell somebody else. Stories spread just because they're good stories. Perhaps they're funny. Perhaps we bask in the attention we get when we pass on a good story. The story of the helium dolls is not only wildly vivid: it chimes with people's expectations or prejudices. Can you see how the same might have been true of stories of Jesus's miracles or his resurrection? Early recruits to the young religion of Christianity might have been especially eager to pass on stories and rumours about Jesus, without checking them for truth.

Think of the distorted legends about 9/11 or the death of Kennedy, and then imagine how even more easily and thoroughly things could have been distorted if there had been no cameras, no newspapers, nothing written down for 30 years after the event. Nothing to go on but word-of-mouth gossip. That was the situation after the death of Jesus. All around the eastern Mediterranean, from Palestine to Rome, there were isolated pockets of Christians of various kinds. Communications between these local groups were poor and infrequent. The gospels were not yet written. They had no New Testament to bind them together. They disagreed on many things, for example whether Christians had to be Jews (and had to be circumcised) or whether Christianity was a whole new religion. Some of Paul's letters show a leader struggling to bring order to this chaos.

An agreed biblical 'canon' – those books agreed as the official list – wasn't finally settled until centuries after

Paul's death. The Bible read by (Protestant) Christians today is a standard canon of 27 books in the New Testament and 39 books in the Old Testament (Roman Catholic and Orthodox Christians have a set of additional books, often called the 'Apocrypha').

Matthew, Mark, Luke and John are the only gospels in the official canon but, as we'll see, plenty of other gospels of Jesus had been written around the same time. The canon was largely fixed in AD 325 by a conference of church leaders called the Council of Nicaea, set up by the Roman Emperor Constantine – the one whose conversion led to Europe becoming Christian. He made Christianity the official religion of the Roman Empire. But for Constantine, you'd probably have been brought up to worship Jupiter, Apollo, Minerva and the other Roman gods. Much later, Christianity was spread across South America by another couple of great empires, the Portuguese Empire (in Brazil) and the Spanish Empire (in the rest of the continent). The widespread presence of Islam in North Africa, the Middle East and the Indian subcontinent is also the result of military conquest.

As I said, Matthew, Mark, Luke and John were only four out of a large number of gospels doing the rounds at the time of the Council of Nicaea. I'll come on to some of the lesser-known gospels in a moment. Any of them could have been included in the canon, but for various reasons none of them made it. Often it was because they were judged heretical, which just means they said things at odds with the 'orthodox' beliefs of council members. Partly it was because they were written slightly more recently than Matthew, Mark, Luke and John. But, as we've seen, even Mark wasn't written early enough to be potentially reliable history.

The favoured four gospels were chosen, in part, for weird reasons which owe more to poetic fancy than to history. Irenaeus, one of those influential figures in the early history of Christianity known as the 'Fathers of the Church', lived a century before the Council of Nicaea. He was convinced that there had to be four gospels, no more and no fewer. He pointed out (as though it mattered) that there are four corners of the earth and four winds. As if that wasn't enough, he also pointed out that the Book of Revelation refers to God's throne being borne by four creatures with four faces. This seems to have been inspired by the Old Testament prophet Ezekiel, who dreamed of four creatures coming out of a whirlwind, each one of which had four faces. Four, four, four, four, you can't get away from four, we obviously have to have four gospels in the canon! I'm sorry to say that's the kind of 'reasoning' that passes for logic in theology.

The Book of Revelation, by the way, wasn't added to the canon until a century later and it's a pity it ever was. Some guy called John had a weird dream one night on an island called Patmos, and he wrote it down. We all have dreams, and many of them are pretty weird. Mine almost always are, but I don't write them down and I certainly don't think they're interesting enough to inflict upon other people. John's dream was weirder than most (almost as though he was on drugs). It has become hugely influential simply because it somehow got itself included in the biblical canon. It's thought to be prophetic and is often quoted by fiery preachers in America. Along with Paul's first letter to the Thessalonians, Revelation is the main inspiration for the idea of 'the Rapture'. It is also the source of the dangerous idea that the longed-for Second Coming of Jesus cannot happen until after the 'Battle of Armageddon'. This belief is why some people in America long for an all-out war involving Israel in the Middle East. They think that war will be 'Armageddon'.

Thousands of people, especially in America since the remarkable popularity of the so-called 'Left Behind' books, sincerely hold the nutty belief that the Rapture is really going to happen. And happen soon. There are even websites that advertise a paid service to look after your pet cat in the event that you are, without warning, hoisted bodily 'up' to heaven. It's a shame people don't realize it was little more than chance which books got included in the canon and which books were . . . left behind!

The long gap between Jesus's death and the gospels being written gives us one reason to doubt that they are a reliable guide to history. Another is that they contradict each other. Although all the gospels agree that Jesus was accompanied by twelve close disciples, they don't agree on who they all were. Matthew and Luke trace the descent of Mary's husband Joseph from King David via two completely different sets of ancestors, 25 of them in the case of Matthew, 41 in Luke. To make matters worse, Jesus was supposed to be born of a virgin mother, so Christians can't use Joseph's descent from David to establish that Jesus was descended from David. There are also discrepancies between the gospels and known historical facts, for example the facts of Roman rulers and their doings.

Yet another problem with taking the gospels as historical truth is their obsession with fulfilling Old Testament prophecies. Especially Matthew. You get the feeling Matthew was quite capable of inventing an incident and writing it into his gospel, simply in order to make a prophecy come true. The most glaring example is his invention of the legend that Mary was a virgin when she gave birth to Jesus. And that's a legend that's really taken on a life of its own. Matthew tells how an angel appeared to Joseph in a dream, reassuring him that his intended wife Mary was pregnant not by another man but by God. (That, by the way, differs from Luke's account, where the angel appears to Mary herself.) Anyway, Matthew goes on, without a hint of shame, to admit to his readers:

> All this took place to fulfil what the Lord had said through the prophet: 'The virgin will be with child and will give birth to a son, and they will call him Immanuel' – which means, 'God with us.'

Perhaps 'shame' was the wrong word for me to use. Matthew, whoever he was, had a different idea of historical truth from ours. For him, fulfilling a prophecy was more important than what actually happened. He wouldn't have understood why I said 'without a hint of shame'.

On the other hand, Matthew totally misunderstood the prophecy. It's in Isaiah, chapter 7. And it's clear from the Book of Isaiah itself – though apparently not to Matthew – that Isaiah was talking not about the distant future, but about the immediate future in his own time. He was talking to the king, Ahaz, about a particular young woman in their presence, who was pregnant even as he spoke.

The word Matthew quoted as 'virgin' was *almah* in Isaiah's Hebrew. *Almah* can mean virgin; but it can also mean 'young woman' – rather like the English word 'maiden', which has both meanings. When Isaiah's Hebrew was translated into Greek in the version of the Old Testament called the Septuagint, which Matthew would have read, *almah* became *parthenos* – which really does mean 'virgin'. A simple translation error spawned the entire worldwide myth of the Blessed Virgin Mary, and the Roman Catholic cult of Mary as a kind of goddess, the 'Queen of Heaven'.

It was the same determination to fulfil prophecies that led both Matthew and Luke to have Jesus born in Bethlehem. Another one of the Old Testament prophets, Micah, had foretold that the Jewish Messiah would be born in Bethlehem, the 'City of David'. John's gospel, reasonably enough, assumes that Jesus was born in Nazareth, which is where his parents lived. John tells of people being surprised that Jesus, if he really was the Messiah, was born in Nazareth. Mark doesn't mention his birth at all. But both Matthew and Luke wanted to fulfil the prophecy of Micah, and both scrambled to find a way to shift Jesus's birthplace from Nazareth to Bethlehem. Unfortunately they did it in two different, contradictory, ways.

Luke's solution to the problem was a tax decreed by the Roman Emperor Augustus. This tax, according to Luke, was accompanied by a census. Luke messed up his dates here, because modern historians know there was no Roman census at the right time to fit the story. But let that pass. In order to be properly counted in the census, everybody had to go to 'his own city'. Although Joseph actually lived in Nazareth, his 'own city', according to Luke, was Bethlehem. Why? Because he was descended in the male line from King David, and David came from Bethlehem. That's ridiculous in itself, by the way. By Luke's own account, David was Joseph's 41-greats-grandfather. How could any law define a man's 'own city' as the city where his 41-greats-grandfather was born? Do you have the faintest idea who your 41-greats-grandfather in the male line was? I doubt that even Queen Elizabeth knows. Anyway, according to Luke, that was why Jesus was born in Bethlehem. His parents moved from Nazareth to be in the birthplace of Joseph's 41-greats-grandfather, for the census.

Matthew's way of fulfilling Micah's prophecy was different. He apparently assumed that Bethlehem was Mary and Joseph's home town, and that was why Jesus was born there. Matthew's problem was how to move them to Nazareth later. So he had the wicked King Herod getting wind of Jesus's birth in Bethlehem. Fearful of a prophesied new 'King of the Jews' who would topple him off his throne, Herod ordered all the boy babies in Bethlehem to be killed. God sent an angel to warn Joseph in a dream, telling him to flee with Mary and Jesus to Egypt. Perhaps you've sung the Christmas carol which goes:

> Herod then with fear was filled:
> A prince, he said, in Jewry!
> All the little boys he killed
> At Bethl'em in his fury.

Mary and Joseph heeded the warning, and didn't return from Egypt until after Herod's death. However, even then they avoided Bethlehem because God warned Joseph, in another dream, that they wouldn't be safe there from Herod's son Archelaus. So they went and lived, instead,

> in a town called Nazareth, that it might be fulfilled which was spoken by the prophets, He shall be called a Nazarene.

Neat solution by Matthew. He got his Jesus character safely to Nazareth, and even managed to score another fulfilled prophecy in the process.

I said I'd return to those extra gospels, about fifty of them, any of which might have been included in the canon along with Matthew, Mark, Luke and John. They were mostly written down in the first couple of centuries AD but, as with the four official gospels, those final written versions were based on older word-of-mouth traditions (complete, presumably, with the usual 'Chinese Whispers' distortions). They include the gospel of Peter, the gospel of Philip, the gospel of Mary Magdalene, the Coptic gospel of Thomas, the infancy gospel of Thomas, the gospel according to the Egyptians and the gospel of Judas Iscariot.

In some cases it's easy to see why they were left out of the canon. Take the gospel of Judas Iscariot, for example. Judas was the arch-villain of the whole Jesus story. He betrayed Jesus to the authorities who then arrested, tried and executed him. According to the gospel of Matthew, his motive was greed: the betrayal earned him 30 pieces of silver. The trouble with Matthew is that, as we've seen, he was obsessed with Old Testament prophecies. Matthew wanted everything that happened to Jesus to be the fulfilment of a prophecy. And we might wonder whether Judas, with his alleged greed motive, was a victim of Matthew's prophet fixation. Here are some clues, which I learned from the biblical historian Bart Ehrman.

The prophet Zechariah (chapter 11, verse 12) was paid 30 pieces of silver. Not a very impressive coincidence. Until you see the next verse from Zechariah:

So they paid me thirty pieces of silver. And the Lord said to me, 'Throw it to the potter' – the handsome price at which they valued me! So I took the thirty pieces of silver and threw them into the house of the Lord to the potter.

Hold 'potter' and 'threw' in your head while we go back to Matthew, chapter 27. Full of remorse, Judas took his 30 pieces of silver to the chief priests and elders.

When Judas, who had betrayed him, saw that Jesus was condemned, he was seized with remorse and returned the thirty silver coins to the chief priests and the elders. 'I have sinned,' he said, 'for I have betrayed innocent blood.' 'What is that to us?' they replied. 'That's your responsibility.' So Judas threw the money into the temple and left. Then he went away and hanged himself. The chief priests picked up the coins and said, 'It is against the law to put this into the treasury, since it is blood money.' So they decided to use the money to buy the potter's field as a burial place for foreigners.

The chief priests didn't want to accept blood money. So instead they used the 30 pieces of silver to buy a field called . . . the Potter's Field. True to form, Matthew rounds off with yet another prophet, this time Jeremiah:

Then what was spoken by Jeremiah the prophet was fulfilled: 'They took the thirty silver coins, the price set on him by the people of Israel, and they used them to buy the potter's field, as the Lord commanded me.'

The rediscovery of the gospel of Judas was one of the most surprising document finds of the twentieth century. People knew that such a gospel had been written, because it was mentioned, and condemned, by early Church Fathers. But everybody thought it was lost, perhaps destroyed as heresy. And then, after 1,700 years, in the late 1970s, it was discovered in a tomb in Egypt. As is usually the case with such finds, it took a while for this priceless document to find its way into the hands of proper scholars capable of taking care of it, and it suffered some damage on the way. It has been carbon-dated to AD 280, plus or minus sixty years.*

The rediscovered document is written in Coptic, an old Egyptian language. But it is thought to be a translation from an earlier, and still lost, Greek text, which was probably nearly as old as the four canonical gospels. Like those four, it was written by somebody other than its named author: so probably not Judas himself. It's mostly a set of conversations between Judas and Jesus. It tells the story of the betrayal, but from Judas's point of view, and it removes much of the blame from him. It suggests that Judas was the only one of the twelve disciples who really understood Jesus's mission. As we shall see in Chapter 4, Christians believe it was God's plan that Jesus should

---

* Carbon dating is a clever scientific technique for dating archaeological specimens; I explained how it works in *The Magic of Reality* (London, Bantam Press, 2011).

be arrested and killed, so that God could forgive humanity's sins. Judas's 'betrayal' was really helping Jesus fulfil God's plan. He was doing Jesus, and God, a favour. If that sounds strange (it does), it gets its strangeness straight from the central idea of Christianity: that Jesus's death was a necessary sacrifice, planned by God. You can see why the Council of Nicaea might not want to include the gospel of Judas in the canon.

For different reasons, it's no surprise that they didn't want the infancy gospel of Thomas either. As usual, nobody knows who wrote it. Contrary to rumour, it wasn't 'Doubting Thomas', the disciple who wanted proof before he believed in Jesus's resurrection (perhaps he should be the patron saint of scientists). This gospel includes amazing stories about Jesus's childhood, a period of his life that's almost completely missing from the official canon. By its account Jesus was a mischievous child, who was not shy of showing off his magic powers. At the age of five, playing by a stream, he took mud from the stream and fashioned it into twelve live sparrows.

A sparrow is made of more than 100 billion cells. Nerve cells, muscle cells, liver cells, blood cells, bone cells and hundreds more different types of cell. Every one of those cells is a miniature machine of mind-blowing complexity. Every one of a sparrow's two thousand feathers is a marvel of delicate architecture. Nobody knew those details in Jesus's time. Even so, you'd think the grown-ups would have been pretty impressed. To make all that out of mud, at a stroke, would be an astounding feat of magic. But no: Joseph gave higher priority to scolding Jesus because he did it on the sabbath day, when Jewish law forbids you to do any work. Some modern Jews won't even flick a light switch on the sabbath. They have a time-switch to do it for them. And there are apartment buildings where, on the sabbath, the lifts stop at every floor – so you don't have to 'work' by pressing a button.

Jesus's response to being scolded was to clap his hands and say: 'Be gone.' Obediently the sparrows flew off, chirping.

According to the infancy gospel, the young Jesus also used his magic powers in less appealing ways. On one occasion he was walking through the village and another child ran up and bumped into his shoulder. Jesus was cross and said to him, 'You will go no further on your way.' That very night the boy fell down dead. Understandably, the grieving parents complained to Joseph and asked him to control Jesus's use of his magic powers. They should have known better: Jesus promptly struck them blind. On an earlier occasion Jesus was annoyed with a boy and cursed him so that his body completely withered up.

It wasn't all bad. When one of his playmates fell off a roof and died, Jesus brought him back to life. He saved a number of people in the same kind of way, and once healed a man who accidentally chopped into his own foot with an axe. One day he was helping his carpenter father, and it turned out that a piece of wood was too short. Well,

Jesus wasn't going to let a little problem like that spoil a good piece of work! He simply lengthened the wood with one of his magic spells.

Nobody thinks the fantastic miracles in the infancy gospel of Thomas really happened. Jesus didn't turn mud into sparrows, didn't kill the boy who bumped into him or blind the boy's parents, or lengthen the piece of wood in the carpenter's shop. Why, then, do people believe the equally far-fetched miracles described in the official, canonical gospels: turning water into wine, walking on water, rising from the dead? Would they have believed the sparrow miracle, or the plank-lengthening miracle, if the infancy gospel had made it into the canon? If not, why not? What's so special about the particular four gospels lucky enough to be chosen for the canon by a bunch of bishops and theologians in Nicaea in AD 325? Why the double standard?

Here's another example of the double standard. Matthew tells us that, at the exact moment of Jesus's death on the cross, the great curtain in the temple in Jerusalem was split down the middle, the earth shook, the tombs broke open and dead people walked the streets. According to the official gospel, then, Jesus wasn't unusual in being resurrected. Only three days before Jesus did it, lots of other people burst out of their graves and walked the streets of Jerusalem. Do Christians really believe that? If not, why not? There's as much reason (or, more to the point, as little reason) to believe it as there is to believe in Jesus's own resurrection. How do believers decide which far-fetched tales to believe and which to ignore?

As I said, most, though not all, historians think Jesus existed. But that isn't saying much. 'Jesus' is the Roman form of the Hebrew name Joshua or Yeshua. It was a common name and wandering preachers were common. So it's not unlikely there was a preacher called Yeshua. There could have been many. What is not believable is that any of them turned water into wine (or mud into sparrows), walked on water (or lengthened a piece of wood), was born to a virgin or rose from the dead. If you want to believe such things, you'd do well to look for much better evidence than is at present available. As the astronomer Carl Sagan said, 'extraordinary claims require extraordinary evidence'. He may have been inspired by Laplace, the renowned French mathematician, who said: 'The weight of evidence for an extraordinary claim must be proportioned to its strangeness.'

The claim that a wandering preacher called Jesus existed is not an extraordinary claim. And the evidence, though slight, is 'proportioned': small evidence for a small claim. Yeshua probably existed. But the claims that his mother was a virgin, and that he rose from the grave, are very extraordinary indeed. So the evidence had better be good. And it isn't.

The great eighteenth-century Scottish philosopher David Hume had something to say about miracles, and I'd like to talk about it because it's important. I'll put it in my own words. If somebody claims to have seen a miracle – makes, for example, the miraculous claim that Jesus rose from his grave, or the miraculous claim that the boy Jesus turned mud into sparrows – there are two possibilities.

Possibility 1: It really happened.

Possibility 2: The witness is mistaken – or is lying, was hallucinating, has been misreported, saw a conjuring trick, etc.

You might say: 'This witness is so reliable, I'd trust him with my life, and there were lots of other witnesses – it would be a *miracle* if he was lying or otherwise mistaken.' But Hume would retort: All well and good, but even if you think Possibility 2 would be a miracle, you'd surely admit that Possibility 1 is even more miraculous. When you have a choice of two possibilities, always choose the less miraculous.

Have you ever seen a really mind-blowing 'magician', a great conjuror? Derren Brown, say, or Jamy Ian Swiss, or David Copperfield, or James Randi or Penn and Teller? It's uncanny, you have an inner voice screaming, 'It's got to be a miracle, there's no way that isn't supernatural.' But then, if the conjuror is honest, he will tell you, calmly and gently, 'No, it's just a trick. I mustn't tell you how it's done, I'd be thrown out of the Magic Circle if I did, but I promise you it's only a trick.'

Not all conjurors are honest, by the way. Some make pots of money by bending spoons with so-called 'psychic powers', and then dishonestly persuading mining companies that those same psychic powers can tell them where to dig. Such fakers have an easy time of it, because their victims are eager to believe in miracles.

Sometimes it's easy to see how the trick is done. I remember a show on British television promoting 'amazing' feats of psychic powers – telepathy and all that. Actually it was nothing but ordinary conjurors fooling a TV presenter called David Frost. David Frost was either very silly, or – more probably – was pretending to be silly for the benefit of the show's ratings. There was a father and son act from Israel, in which the son claimed to read his father's thoughts by telepathy. The father looked at a secret number and sent 'waves of thought' to his son at the other side of the stage, who correctly 'read his thoughts'. The father put on a great act of concentration and then shouted out something like 'Have you got it, son?' – at which the son would shout 'Five!' The audience burst into wild applause, egged on by the foolish host: 'Amazing! Uncanny! Deeply mysterious! Telepathy is proved!'

Have you got it? Let me give you a hint. If the secret number had been eight, the father would have shouted something like 'Do you think you can do it, son?' If the secret number was three, it would have been 'Got it, son?' If the number was four, 'Got it yet, son?' But my point is that, even if the conjuror is a really good conjuror (unlike that father and son team) and you simply cannot begin to guess how the trick is done, it's still a trick. There's no reason to resort to 'it must be a miracle'. Think like Hume.

Let's apply Hume's reasoning to some famous conjuring tricks, renaming the two 'Possibilities' as 'Miracles'.

> **Miracle 1:** The conjuror really did saw the woman in half. Penn and Teller really did catch the bullets from each other's pistols in their teeth. David Copperfield really did make the Eiffel Tower disappear. James Randi really did penetrate a patient's abdomen with his bare hands and haul out the guts.

> **Miracle 2:** Your eyes deceived you, even though you were watching the conjuror's every move like a hawk, so it would seem 'miraculous' for you to miss anything.

I think you have to agree that 'Miracle' 2, however much you want to protest, is less of a miracle. You have to prefer the lesser miracle and conclude, with Hume, that Miracle 1 never happened. You were deceived.

Sometimes Miracle 1, the allegedly real miracle, seems to be confirmed by the sheer number of witnesses. Perhaps the most famous example is the Apparition of Our Lady of Fatima.

In 1917, at Fatima in Portugal, three children claimed to have seen a vision of the Virgin Mary. One of them, Lucia, said Mary had spoken to her and had promised to return to the same spot on the 13th of each month until October, when she would do a miracle to prove who she was. Rumours spread all around Portugal. And on 13 October a huge crowd of seventy thousand gathered to witness the miracle. Sure enough, according to witnesses, it happened. The Virgin Mary appeared to Lucia (nobody else), who pointed excitedly towards the sun. Then—

> the sun seemed to tear itself from the heavens and come crashing down upon the horrified multitude . . . Just when it seemed that the ball of fire would fall upon and destroy them, the miracle ceased, and the sun resumed its normal place in the sky, shining forth as peacefully as ever.

Roman Catholics took the story seriously (a lot of them still do). They declared it an official miracle. Pope John Paul II survived an assassination attempt in 1981. He believed he was saved by 'Our Lady of Fatima' who 'guided the bullet' so it didn't kill him. Not just 'Our Lady' but specifically 'Our Lady of Fatima'. Does this mean Catholics believe in lots of different 'Our Ladies'? Are they even more polytheistic than I suggested in Chapter 1? Not just one Mary but lots of Marys, one for each appearance in some hillside or cave or grotto.

In 2017 Bishop Dominick Lagonegro, Roman Catholic Auxiliary Bishop of New York, preached a sermon in which he quoted his aunt, who had been an eye-witness at Fatima. By her account, the sun

> went up and down and turned back and forth, almost as if it were dancing. 'Who else but the Blessed Mother could make the sun dance,' [Bishop Lagonegro] laughed. But then it got big and 'started coming to the earth,' the bishop continued. 'My aunt recalled that "it looked as if everyone's clothes were bright yellow from the sun". It continued to fall to the earth for a few minutes,' he said, telling her story, 'and then stopped', going back into its orbit.

Its 'orbit'? What 'orbit' would that be? And it 'continued to fall to the earth for a few minutes'. For a few *minutes*! Let's do a Hume on the case.

> **Miracle 1:** The sun really did move about the sky and then start to come crashing down towards the crowd, moving perceptibly towards them for several minutes.

> **Miracle 2:** Seventy thousand witnesses were mistaken, or lied, or were misreported.

Miracle 2 really does seem like a miracle, doesn't it? Seventy thousand people all had the same hallucination at the same time? Or all told the same lie? Surely that would be a gigantic miracle? So it would seem. But consider the alternative, Miracle 1. If the sun really had moved, wouldn't it have been seen by everybody on the daylight side of the world? Not just the people gathered outside a single village in Portugal? And if it really had moved (or the Earth had moved so that it looked as though the sun had moved), it would have been a catastrophe which would have destroyed the world if not all the other planets too. Especially if it 'fell' for 'a few *minutes*'!

So, following Hume, we choose the lesser miracle and conclude that the famous miracle of Fatima never happened.

Actually, I was bending over backwards to make 'Miracle' 2 seem more miraculous than it really was. Were there really seventy thousand people there? What is the historical evidence for such a large number? In our own time such numbers are often exaggerated. Donald Trump claimed that one and a half million people attended his inauguration as President. Photographic evidence shows that to be a massive exaggeration. Even if seventy thousand did converge on Fatima in October 1917, how many of them really claimed to see the sun move? Maybe only a few did, and the number was inflated by the Chinese Whispers effect. If you stare at the sun, as Lucia told them to (don't try it, by the way, it's bad for your eyesight), you might well hallucinate a slight movement. Then the size of that movement, as well as the number who saw it, could be exaggerated by the Chinese Whispers effect.

But the important point is that we don't need to bother with those considerations. Even if a full seventy thousand people really did claim to see the sun move and come crashing down, we know for certain it didn't really happen because the planet wasn't destroyed and nobody outside Fatima saw it move. The alleged miracle certainly never happened and the Roman Catholic Church was very silly to grant it official authentication.

Incidentally, a similar miracle is reported in the Book of Joshua. Maybe this was what inspired Lucia to invent hers. The Israelite leader Joshua was having one of his many battles with rival tribes and he needed a bit more time to secure his victory. What to do? The obvious solution! You could talk to God directly in those days. All Joshua had to do was ask God to postpone nightfall by making the sun stand still in the sky. God obliged and the sun stood still, providing Joshua with the extra-long day he needed to win his battle. Obviously this miracle never actually happened. No serious scholar thinks it did. But there are fundamentalist Christians who yearn to believe that every single word of the Bible is literally true. And you can find fundamentalist websites that desperately twist and turn to find ways to make the miracle of Joshua's long day true.

The Book of Joshua, of course, is one of the books of the Old Testament. We now turn to the Old Testament itself, and ask whether any of its stories are true.

· 3 ·

Myths and
how they start

In Chapter 2 I talked mainly about the New Testament. Dealing with more recent times than the Old, it's the Bible's best shot at being history. I won't spend long on the Old Testament. It takes us further into the shadowy realms of myth and legend, and biblical scholars don't take it seriously as history. But myths are interesting and important in their own right, and this chapter will use the Old Testament as a starting point to take a look at myths and how they start.

Abraham was the original patriarch of the Jewish people and founder of the three main monotheistic religions in the world today – Judaism, Christianity and Islam. But did he really exist? As with Achilles and Hercules, as with Robin Hood and King Arthur, it's impossible to know, and there's no positive reason to think he did. On the other hand, Abraham's existence is not an extraordinary claim requiring extraordinary evidence. Unlike Joshua's long day or Jesus's resurrection, or Jonah living three days in the belly of a big fish, Abraham's existence – or not – is no big deal. There just isn't any evidence, one way or the other. Same with King David, another great hero of Jewish history. David made no impact either on archaeology or on written history outside the Bible. This suggests that, if he existed at all, he was probably a minor local chieftain rather than the great king of legend and song.

Talking of song, the Song of 'Solomon' (also known as the Song of Songs, which is a better title, for it certainly wasn't written by King Solomon) is the only sexy book in the Bible. It's pretty surprising the Council of Nicaea allowed it in the official canon. Here's something rather funny about it. The King James Bible, the most famous English translation, has commentary lines at the top of each page. The Song is a wonderful poetic expression of sexual love between a woman and a man. But what does the Christian commentary say at the top of the page? 'The mutual love of Christ and his church.' Priceless. And utterly typical of the way theologians think: ignore what is actually being said, and pretend it was all intended to be a symbol or a metaphor.

There's some beautiful English writing in the King James Bible. Ecclesiastes is at least as good as the Song of Songs, although its poetry is bleak and world-weary. If you read nothing else in the Bible, I recommend those two books, Ecclesiastes and the Song of Songs. But make sure you read the King James version. Translations into modern English just don't work. As poetry, that is. They do work if you want to get a truer idea of what the original Hebrew said. And that's likely to help you understand things that religious teachers might prefer you not to understand! If you don't know what I mean by that, wait till Chapter 4.

Those two favourite books of mine, Ecclesiastes and the Song of Solomon, don't pretend to be history. Other books of the Old Testament do, for example Genesis, Exodus, Kings and Chronicles. Genesis, Exodus, Leviticus, Numbers and Deuteronomy are called the Pentateuch by Christians, and the Torah by Jews. Moses is traditionally supposed to have written them, but no serious scholar thinks he did. As with the stories of Robin Hood and his Merry Men, or King Arthur and his Knights of the Round Table, there may be some obscure fragments of truth buried in the Pentateuch, but there's nothing you could call real history.

The great ancestral myth of the Jewish people is their captivity in Egypt and their heroic escape to the Promised Land. That was Israel, the land flowing with milk and honey, the land God said should be theirs and for which they fought the tribes who already lived there. The Bible obsessively repeats this legend. And the leader who is supposed to have led the Jews out of Egypt to the promised land was Moses, the same Moses who, they believed, was the author of the first five books of the Bible.

You would think that such a big event as the enslavement of an entire nation, and its mass migration generations later, would have left traces in the archaeological record and in the written histories of Egypt. Unfortunately there is no evidence of either kind. No evidence of anything like a Jewish captivity in Egypt. It probably never happened, although the legend is burned deep into Jewish culture. When the Bible mentions either God or Moses, their name is likely to be followed by 'who brought you out of Egypt' or some equivalent phrase.

The alleged escape from Egypt is remembered by Jews every year in the Feast of the Passover. Fiction or fact, it's not a pretty story. God wanted the Egyptian king, the Pharaoh, to set the Israelite slaves free. You might have thought it would be within God's powers to change Pharaoh's mind miraculously. He deliberately did the exact reverse, as we shall see. But first he put pressure on Pharaoh by sending a series of ten plagues to Egypt. Each plague was nastier than the last, until eventually Pharaoh gave up and freed the slaves. Among them were a plague of frogs, a plague of painful boils, a plague of locusts, and darkness for three days. The final plague was the clincher, and it's this one the Passover commemorates. God killed the eldest child in every Egyptian household, but 'passed over' the houses of Jews, sparing their children. The Israelites were told to paint their doorposts with lambs' blood, so the angel of death could tell which houses to avoid on the child-slaughtering spree. You'd think that God, being all-wise and all-knowing, might have been able to tell which house was which. But perhaps the author thought the lambs' blood would add a nice splash of colour to the story. Anyway, that was the legendary Passover event which is still celebrated by Jews everywhere.

Actually, Pharaoh had been on the point of giving up and letting the Israelites go earlier, and that would have been nice because all those innocent children would have been saved. But God deliberately used his magic powers to make Pharaoh obstinate, so that God could send some more plagues, as 'signs' to show the Egyptians who was boss. Here's what God said to Moses:

> But I will harden Pharaoh's heart, and though I multiply my miraculous signs and wonders in Egypt, he will not listen to you. Then I will lay my hand on Egypt and with mighty acts of judgment I will bring out my divisions, my people the Israelites. And the Egyptians will know that I am the Lord when I stretch out my hand against Egypt and bring the Israelites out of it. (Exodus 7: 2–3)

Poor Pharaoh. God 'hardened his heart' in order to make him refuse to free the Israelites, specifically so that God could do his Passover trick. God even told Moses in advance that he would make Pharaoh say no. And the blameless firstborn children of the Egyptians were all killed as a result. By God. As I said, it's not a pretty story and we can be thankful it never really happened.

Much more authentic than the alleged captivity of the Jews in Egypt is their later captivity in Babylon. There's plenty of evidence for that. In 605 BC, the Babylonian King Nebuchadnezzar besieged Jerusalem and carried off many of the Jews. About 60 years later, Babylon itself was conquered by the Persian king Cyrus the Great. Cyrus permitted the Jews to return home, which some of them did. It was during or around the time of the Babylonian exile that most of the Old Testament books were written. So, if you thought the stories of Moses or David, Noah or Adam, were written by people with up-to-date knowledge of what allegedly happened, think again. Most of the apparent history in the Old Testament was written much more recently – between 600 and 500 BC, many centuries after the events they purport to describe.

We get clues to when the Old Testament was actually written from anachronisms in the text. An anachronism is something that crops up in the wrong time, say when an actor in a costume drama about ancient Rome forgets to take his wristwatch off. Well, here's a nice anachronism in the book of Genesis. Genesis says Abraham owned camels. But archaeological evidence shows that the camel was not domesticated until many centuries after Abraham is supposed to have died. Camels had, though, been domesticated by the time of the captivity in Babylon, which is when the book of Genesis was actually written.

What, then, can we say about the myths from the beginning of Genesis? Adam and Eve? Or Noah's Ark? The Noah story comes directly from a Babylonian myth, the legend of Utnapishtim – which isn't surprising, since Genesis was written during the Babylonian captivity. The Utnapishtim story in turn comes from the Sumerian Epic of Gilgamesh. Arguably the world's oldest work of literature, it was written two thousand years earlier than the Noah story. The Sumerians were polytheists. Their flood legend says the gods couldn't get to sleep because humans made so much noise. Fed up with the racket, the gods decided to drown everybody in a great flood. But one of the gods, the water god Enki, took pity on a man called Utnapishtim (Ziusudra in an older version) and warned him to build a huge boat, to be called 'The Preserver of Life'. The rest of the story is pretty much the same as the Noah version: animals of every kind taken on board, a dove, a swallow and a raven released from the ark to see if there was any land coming up, and so on, including the spectacular rainbow finish. It was another god, Ishtar, who put up the rainbow as a sign that there would be no more catastrophic floods.

Greek mythology has a related story. Zeus, the king of the gods, furiously decided to put an end to humankind. He flooded the world and drowned everybody. Everybody, that is, except one couple, Deucalion and his wife Pyrrha. They survived in a floating chest which eventually came to rest on Mount Parnassus. All around the world, there are similar myths of a great flood in which only one family survived. In the Aztec legend from ancient Mexico, the sole survivors, Coxcox and his wife, floated in a hollow tree trunk and finally, like Noah, landed on a mountaintop and descended to repopulate the world.

In blissful ignorance of the story's polytheistic roots in Babylon, Bible-believing Christians in Kentucky raised the (tax-free) money to build a gigantic wooden Noah's

Ark, which people pay to visit. You'd think they might have given a bit more thought to the story. If the tale of Noah were true, the places where we find each kind of animal should show a pattern of spreading out from the spot where the biblical Ark finally came to rest when the flood subsided – Mount Ararat in Turkey. Instead, what we actually see is that each continent and island has its own unique animals: marsupials in Australia, South America and New Guinea, anteaters and sloths in South America, lemurs in Madagascar. What were those people in Kentucky thinking? Did they imagine that Mr and Mrs Kangaroo came bounding out of the ark and hopped all the way to Australia without having any children on the way? Plus Mr and Mrs Wombat, Mr and Mrs Tasmanian Wolf, Mr and Mrs Tasmanian Devil, Mr and Mrs Bilby and lots of other marsupials not found anywhere except Australia. Mr and Mrs Lemur – all 101 pairs of them – made a beeline for Madagascar and nowhere else! And did Mr and Mrs Sloth crawl – oh, so slowly – all the way to South America? In fact, of course, all the animals, and their fossils, are exactly where they should be according to the principles of evolution. This was one of the main pieces of evidence Charles Darwin used. Ancestral marsupial mammals evolved separately in Australia over millions of years, branching into lots of different marsupials – kangaroos, koalas, opossums, quokkas, phalangers and so on. A different set of mammals evolved in South America, branching, over millions of years, into sloths, anteaters, armadillos and their kind. Yet another set in Africa. Yet another set, including all the lemurs, in Madagascar. And so on.

The stories of Adam and Eve, and of Noah and his Ark, are not history, and no educated theologian thinks they are. Like countless such stories from all over the world, they are 'myths'. There's nothing wrong with myths. Some are beautiful and most are interesting, but they aren't history. Unfortunately, many uneducated people, especially in America and the Islamic world, think they are. All peoples have myths. The two I've just been talking about are Jewish myths, which have become extremely well known throughout the world simply because they happened to be gathered into the sacred canons of Judaism, Christianity and Islam.

It's seldom clear how an ancient myth got started. Perhaps there was an original story about something that did actually happen, say a daring deed by some local hero like Achilles or Robin Hood. Maybe an imaginative storyteller entertained people round the camp fire with a yarn – which might have been either a garbled version of something that once happened, or a piece of fiction made up just for fun, perhaps like the tale of Sinbad the Sailor. Such a storyteller might make use of characters from earlier myths that would have been already well known to his audience: figures like Hercules, Achilles, Apollo, Theseus. Or, coming up to our own time, like Brer Rabbit or Superman or Spider-Man. What's more, the storyteller might not have thought of his stories as pure fiction for entertainment. He might have intended them as moral tales. Like Jesus's parable of the Good Samaritan. Or like Aesop's Fables.

Myths often have a dreamlike quality, and sometimes the original inventor of the story may have been recounting a dream. Throughout history, lots of people have believed their dreams were filled with meaning. Dreams have been thought to foretell the future. Australian aborigines trace their mythology from a mysterious dawn age in the ancestral past, which they call the Dreamtime.

However a story starts, whether in truth or fiction, parable or dream, the Chinese Whispers effect will see to it that it changes as it's repeated and re-repeated down the generations. Noble deeds become exaggerated, eventually often to superhuman levels. Sometimes the names get altered, as when the Utnapishtim character in the Sumerian legend became the Noah character in the Hebrew retelling. All kinds of details change. Successive storytellers 'improve' the story, changing details to make it funnier. Or to make it fit with their previous beliefs or wishful thinking. Or simply to make events in the story more typical of an already well-loved character. So, by the time the story is finally written down, little of the original survives. It's become a myth.

The development of a myth can be very rapid, as we know from those fascinating cases that have started in our own time so that we've actually been able to watch their birth and development. There are many myths about Elvis Presley being seen alive, which might make you think twice about the similar stories of Jesus's resurrection.

My favourite example of a modern myth is the 'cargo cults' of New Guinea and various Melanesian islands in the Pacific. During the Second World War, many islands were occupied by Japanese, American, British or Australian troops. These military outposts were richly supplied with goods – food, fridges, radios, telephones, cars and so on. Something similar had been going on since the nineteenth century, with supplies brought in for colonial administrators, missionaries and so on. But the scale of the wartime deliveries especially dazzled the islanders. No foreigner was ever seen growing crops, or making cars or fridges, or doing almost anything useful. And yet those wonderful things kept arriving, dropping out of the sky. Literally out of the sky during the war, because they came in big cargo planes. It seemed obvious to the islanders that all that lovely cargo must come from the gods, or from the ancestors (who were worshipped as gods). And since the invaders never did any useful work to get stuff, the things they did do must be religious ceremonies. They must be designed to please the cargo gods and persuade them to rain yet more cargo down from heaven. So the islanders tried to imitate these ceremonies, thinking this would please the cargo gods.

How did they do that? Well, it was clear that the airport must be some kind of sacred, holy place, because that's where the cargo planes homed in. So the islanders decided to make their own 'airport' in a forest clearing, complete with dummy control tower, dummy radio masts and dummy planes on the dummy runway. After the war, when the military outposts had departed and the cargo stopped arriving from the sky, the islanders hoped for a 'second coming'. They redoubled their efforts to please the cargo gods and bring back the lost but remembered age of glorious plenty.

Cargo cults sprang up dozens of times independently, on lots of islands widely separated from each other. Some of them are still going strong. On the island of Tanna (Vanuatu), the related cult of 'John Frum' still exists. John Frum is a mythical, messiah-like figure who, the islanders believe, will one day return to take care of his people. Like Jesus. The name seems to come from an American soldier known as 'John from America' (in American English 'from' sounds like 'frum', rhyming with 'come'). Another version of the cult worships 'Tom Navy'. In each case, the name may be grafted on to a personality derived from an older tribal god – just as when 'Utnapishtim' became 'Noah'.

Yet another cult, also on Tanna, worships Prince Philip as a god. Not exactly cargo in this case, but a tall, handsome naval officer who must have looked pretty dazzling in his white uniform, and sufficiently god-like to be cheered by crowds wherever he went. That seems to have kick-started the Chinese Whispers process. The

Prince Philip myth has grown ever since 1974 when he visited the island, and some inhabitants are still, in 2018, looking forward to his Second Coming.

These modern religious cults give us a good idea of how easily myths can arise. Perhaps you've seen Monty Python's film *Life of Brian*? The hero, Brian, is unfortunately mistaken for the Messiah. Running frantically away from the adoring crowds, he drops a gourd and also loses one of his sandals. Almost immediately there is a 'schism' with the worshippers splitting into two rival groups. One group follows the sacred sandal, the other group the sacred gourd. Do see the film if you get the chance – it is very funny indeed, and a perfect satire on the way religions get started.

David Attenborough, one of my favourite people (surely one of everybody's favourite people), tells of a conversation he had on Tanna with a John Frum worshipper called Sam. He pointed out to Sam that after 19 years the second coming of John Frum had still not happened.

> Sam lifted his eyes from the ground and looked at me. 'If you can wait two thousand years for Jesus Christ to come an' 'e no come, then I can wait more than nineteen years for John.'

Sam had a point (although he was wrong to assume David Attenborough is a believing Christian). The early Christians believed Jesus's Second Coming would happen during their own lifetimes, and his own words, as quoted in the gospels, suggest that Jesus – or at least, the people who wrote his teachings down – thought so too.

Mormonism is another relatively recent cult which, unlike the John Frum or cargo cults, or the 'Elvis is Risen' cult, has spread all over the world and become rich and powerful. The founder was a man from New York State called Joseph Smith. He claimed that in 1823 an angel called Moroni told him where to dig up some golden plates which had ancient writing on them. Smith said he did so, and translated the writing from an old Egyptian language into English. He did this with the aid of a magic stone in a magic hat. When he looked in the hat, the stone revealed to him the meaning of the words. He published his English 'translation' in 1830. Weirdly, the English was not the English of his own time but the English of more than two centuries earlier, the English of the King James Bible. Mark Twain joked that if you cut out every repetition of 'It came to pass', the Book of Mormon would shrink to a pamphlet.

Why? What did Smith think he was playing at? Did he think God spoke English? And sixteenth-century English at that? It reminds me of the story (perhaps false, but very 'spreadable', like the story of the helium-filled dolls) of an ex-Governor of Texas called Miriam A. Ferguson. Disliking the idea of Spanish being made an official language in Texas, she is alleged to have said: 'If English was good enough for Jesus, it's good enough for me.'

You'd think Joseph Smith's use of archaic English would have been enough to arouse people's suspicions that he was a fake. That plus the fact that a court had earlier found him guilty of fraud. Nevertheless, he soon attracted followers, and now he has millions. Not long after Smith was murdered in 1844, his cult grew into a major new religion, under a charismatic leader called Brigham Young. Moses-like (you see how myths borrow from earlier myths), Brigham Young led his followers on a wandering pilgrimage to find a promised land. It turned out to be the state of Utah. Today they pretty much run the state. And Mormonism has now spread around the world under the name 'Church of Latter Day Saints' or 'LDS'. There's a colossal Mormon temple in Salt Lake City and at least a hundred more great temples around America and the world. Mormonism is no longer a local cult like the John Frum cult of Vanuatu. Mormons include prosperous leaders of American industry, men in suits with university degrees, one man who almost became US President. Mormons are expected to give 10 per cent of their income to the church, which as a result has become fabulously wealthy – as you can see if you look at those amazing temples.

Yet these prosperous Mormon gentlemen believe things which are positively known, from scientific evidence, to be absurd: complete and utter made-up nonsense. For example, the Book of Mormon explains in detail that Native Americans are descended from Israelites who migrated to North America around 600 BC. As if it weren't obvious, DNA evidence conclusively shows this to be false. Once again, you might think that would be enough to show the Mormons that Smith was a charlatan. But not a bit of it.

It gets worse. Some years after producing the Book of Mormon, Smith claimed to have translated some ancient Egyptian documents that had been bought by a collector after they were discovered near Thebes in Egypt. Smith published his 'translation' as the 'Book of Abraham' in 1842, claiming it was a description of Abraham's life and journey to Egypt. There's lots of detail about Abraham's early life and about Egyptian history and astronomy, pages and pages of it. In 1880, Smith's Book of Abraham was officially 'canonized' by the Mormon Church.

Experts on Egyptian hieroglyphics suspected that Smith's 'translation' was a fake. In the words of a 1912 letter by a curator in the Metropolitan Museum of New York, the Book of Abraham was 'a pure fabrication ... a farrago of nonsense from beginning to end'. But it was still possible for devout Mormons to keep faith with it, for the original papyruses were supposed to have been lost when the Chicago museum that housed them caught fire in 1871. Unfortunately for Joseph Smith, not all the papyruses were destroyed. Some of them were rediscovered in 1966. By this time, scholars understood the language in which the documents were written. When they were properly translated, by both Mormon and non-Mormon scholars who actually knew the language, it turned out they were about something completely different. Nothing to do with Abraham at all. Joseph Smith's 'translation' was an elaborate, and obviously deliberate, hoax.

So, we positively know that Smith's Book of Abraham was a fake translation of manuscripts that really existed. Isn't it rather likely that his earlier 'translation' of the Book of Mormon, using a magic stone in a magic hat, and working from 'golden plates' which mysteriously 'disappeared' so that nobody else could see them, was also a fake? You might think the Mormons would have got the point. But even Smith's obviously dishonest faking of the 'Book of Abraham' wasn't enough to shake the faith of believers.

I suspect that this shows the amazing power of childhood indoctrination. People who are brought up in a religion have great difficulty shaking it off. And then they pass it on to the next generation. And so on. The Church of Latter Day Saints is now one of the fastest-growing religions in the world. Think about that, and perhaps you can see how, in an earlier age when there were no newspapers, no internet, no books, nothing but word-of-mouth gossip for decades after Jesus's death, the cult of Christ was able to take off – virgin birth, miracles, resurrection, ascension into heaven and all.

Unlike the myths of Mormon and John Frum, the Old Testament myths like the Garden of Eden were invented too long ago for us to know how they started. Every tribe has its origin myth – not surprisingly, since people are naturally curious about where they came from, where all the animals came from, how the world, the sun, the moon and the stars came into being. The story of the Garden of Eden is the origin myth of the Jews. Out of all the thousands of origin myths from around the world, it was the Jewish origin myth that happened to get into the Christian Bible – simply because of the twin historical accidents of Jesus being a Jew and the Emperor Constantine's conversion to Christianity. Unlike the Noah story, the Adam and Eve myth might not come from a Babylonian source. Funnily enough, it has similarities with the origin myth of the Pygmies, tiny people who live in the Central African forests.

You'll remember that, in the Jewish myth, Adam was created from 'the dust of the ground'. God 'breathed into his nostrils the breath of life; and man became a living soul'. Rather like a gardener, God then grew Eve as a sort of cutting, from one of Adam's ribs. Incidentally, you'd be amazed at how many people seriously think – on the basis of this myth – that men have one rib missing!

Adam and Eve were placed in a lovely garden, the Garden of Eden. God told them they were free to eat anything they liked in the garden, with one important exception. One particular tree in the middle of the garden, the Tree of Knowledge of Good and Evil, was strictly out of bounds. They must on no account eat its fruit. That was fine for a while. But then a talking snake sidled up to

Eve and persuaded her to eat the forbidden fruit of the Tree of Knowledge. She did, and she persuaded Adam to try it too. Alas! Immediately they were filled with forbidden knowledge, including the fact that they were naked. Shy of their nakedness, they made themselves aprons out of leaves. This gave the game away to God, 'walking in the garden in the cool of the day' (lovely phrase). He realized that they must have eaten the dreaded fruit. He was furious. Poor Adam and Eve were banished for ever from the beautiful garden. Adam and his male descendants were condemned to back-breaking labour all their lives. Eve and her female descendants were condemned to the pains of childbirth. And the snake and its descendants were condemned to slither along the ground without any legs (and presumably to losing the power of speech).

Now compare that Jewish origin myth with this one from the Pygmies. The resemblance was pointed out by a Belgian anthropologist who lived among the Pygmies of the Ituri forest, studied their language and translated various similar versions of their origin myth. Here's one version.

> One fine day in heaven, God told his chief helper to make the first man. The angel of the moon descended. He modelled the first man from earth, wrapped a skin around the earth, poured blood into the skin, and punched holes for the nostrils, eyes, ears and mouth. He made another hole in the first man's bottom, and put all the organs in his insides. Then he breathed his own

vital force into the little earthen statue. He entered into the body. It moved . . . It sat up . . . It stood up . . . It walked. It was Efé, the first man and father of all who came after.

God said to Efé, 'Beget children to people my forest. I shall give them everything they need to be happy. They will never have to work. They will be lords of the earth. They will live forever. There is only one thing I forbid them. Now – listen well – give my words to your children, and tell them to transmit this commandment to every generation. The tahu tree is absolutely forbidden to man. You must never, for any reason, violate this law.'

Efé obeyed these instructions. He, and his children, never went near the tree. Many years passed. Then God called to Efé, 'Come up to heaven. I need your help!' So Efé went up to the sky. After he left, the ancestors lived in accordance with his laws and teachings for a long, long time. Then, one terrible day, a pregnant woman said to her husband, 'Darling, I want to eat the fruit of the tahu tree.' He said, 'You know that is wrong.' She said, 'Why?' He said, 'It is against the law.' She said, 'That is a silly old law. Which do you care about more – me, or some silly old law?'

They argued and argued. Finally, he gave in. His heart pounded with fear as he sneaked into the deep, deep forest. Closer and closer he came. There it was – the forbidden tree of God. The sinner picked a tahu fruit. He peeled the tahu fruit. He hid the peel under a pile of leaves. Then he returned to camp and gave the fruit to his wife. She tasted it.

She urged her husband to taste it. He did. All of the other Pygmies had a bit. Everyone ate the forbidden fruit, and everyone thought that God would never find out.

Meanwhile, the angel of the moon watched from on high. He rushed a message to his master: 'The people have eaten the fruit of the tahu tree!' God was infuriated. 'You have disobeyed my orders,' he said to the ancestors. 'For this you will die!'

Well, what do you think? Is it coincidence? The resemblance is not close enough to be sure. Maybe there are patterns deeply buried in the human unconscious mind that pop out in the form of myths. The famous Swiss psychologist C. G. Jung called these unconscious patterns 'archetypes'. Jung might suggest that Forbidden Fruit is a universal human archetype which lurked in both Pygmy minds and Jewish minds, and independently inspired their two origin myths. Perhaps we need to add Jung's archetypes to our list of how myths around the world get started. Could the widespread myth of a great worldwide flood also be a Jungian archetype?

Another possibility, which may have already occurred to you, is that Pygmy mythology is not pure Pygmy in origin. Could it have been contaminated at some stage by Christian missionaries? The missionaries would have taught the Adam and Eve story to the Pygmies. Then, after generations of Chinese Whispers garbling in the deep forest, the biblical idea of forbidden fruit got incorporated into the Pygmies' own origin myth. I think that's pretty likely. Against this, Jean-Pierre Hallet, the Belgian anthropologist who translated the myth (an amazing character, by the way – try Googling his name plus 'badass'), was convinced that the influence went the other way. He thought the legend of the forbidden fruit originated with the Pygmies and spread to the Middle East via Egypt. If either of these theories is right, the differences between the two myths demonstrate yet again the power of the Chinese Whispers effect as one myth morphed into the other.

Many tribal myths, including the Adam and Eve myth, have a poetic beauty. But there's one thing I unfortunately have to repeat, because too many people don't realize it: they are not true. They aren't history. Most of them aren't even remotely based on history. We tend to think the United States is an advanced, well-educated country. And so it is, in part. Yet it is an astonishing fact that nearly half the people in that great country believe literally in the story of Adam and Eve. Luckily the other half is there too, and they have made the United States the greatest scientific power in the history of the world. You have to wonder how much further ahead they would be if they weren't held back by the scientifically ignorant half who believe every word of the Bible is literally true.

No educated person today thinks either the Adam and Eve myth or the Noah's Ark myth is literally true. Plenty of people do, however, believe in the Jesus myths (like Jesus rising from the grave), the Islamic myths (like Mohammed riding a winged horse) or the Mormon myths (like Joseph Smith translating golden tablets). Do you think they are right to do so? Is there good reason to believe those – any more than the myth of the Garden of Eden? Or Noah? Or John Frum and the cargo cults? And, if you believe the myths of your own faith, whichever faith you happen to have been brought up in, why are those myths any more likely to be true than the myths of other faiths, believed equally fervently by other people?

So, we've dealt with the Bible as history. It mostly isn't. And we've dealt with the Bible as myth. Much of it is, and there's nothing wrong with that. Myths are rightly valued. But there's nothing to single out the biblical myths as any more valuable than the myths of the Vikings, the Greeks, the Egyptians, the Polynesian islanders, the Australian aborigines, or any of the countless tribes of Africa, Asia or the Americas. The Bible has one more important claim, however. It's called 'The Good Book', a book of moral wisdom, a book that will help us to lead a good life. Many people, especially in America, go so far as to believe you can't be a good person without it.

Does the Bible deserve its virtuous reputation as the Good Book? You may like to decide after reading the next chapter.

· 4 ·

The Good Book?

'The animals went in two by two.' We love the story of Noah's Ark. Mr and Mrs Giraffe, Mr and Mrs Elephant, Mr and Mrs Penguin and all the other couples, patiently walking up the gangway into the great wooden ship, welcomed by a beaming Mr and Mrs Noah. Sweet. But wait; why was there a worldwide flood in the first place? God was angry with the sinfulness of humankind. All except Noah, who 'found favour in the eyes of the Lord'. So God decided to drown every man, woman and child, plus all the animals except one pair of each kind. Not so sweet after all?

Whether or not we think God is an entirely fictional character, we can still judge whether he is good or bad, just as we might judge Lord Voldemort or Darth Vader or Long John Silver or Professor Moriarty or Goldfinger or Cruella de Vil. So throughout this chapter, when I say 'God did so-and-so' I mean 'the Bible says that God did so-and-so', and from these accounts we can judge if the God *character* is a nice character, whether the stories about him are fact or fiction. I shall do so, and you will no doubt feel free to decide for yourself whether you think it's still possible to love God in spite of everything. As a man called Job did, in the following story from the Bible.

Job was a very good, righteous man who loved God. This pleased God so much that he had a sort of bet with Satan about Job. Satan thought Job was good and well-behaved and loved God only because he was fortunate – rich and healthy, with a nice wife and ten lovely children. God bet Satan that Job would go on being good and go on loving and worshipping him, even if he lost all his good fortune. God gave Satan permission to test Job by depriving him of everything. And Satan duly set about it. Poor Job! His cattle and sheep all died, his servants were all killed, his camels were stolen, his house blew down in a gale and all his ten children died. But God won the argument because, even in the face of such provocation, Job never became cross with God, and refused to stop loving and worshipping him.

Satan still wouldn't admit defeat, though, so God gave him permission to test Job even further. This time Satan covered Job's whole body with boils, like the boils God inflicted on the Egyptians (caused by bacteria, as we now know, though the author of the book of Job didn't – and presumably God and Satan did). Still Job's faith held firm. He didn't stop loving God. So God finally rewarded Job by curing the boils and giving him lots more wealth. His wife had lots more children. And they all lived happily ever after. Pity about the ten dead children and all the other people who'd been killed because of the bet but – as people often say – you can't make an omelette without breaking eggs.

Like the Noah myth, it's just a story, it didn't happen. As with most books in the Bible, we don't know who wrote the book of Job. And we don't know whether the author himself (it probably was a himself rather than a herself) thought there was a real man called Job. He could have been using fiction to teach a lesson. This is quite likely, because the bulk of the book of Job consists of lengthy dialogues between Job and his friends (known as 'Job's comforters') about moral questions and duty to God. But whatever the author's intention, huge numbers of devout Christians and Jews still think it's a real story about a real, suffering man called Job. Devout Muslims, too, for the story of Job is in the Quran. So is the story of Noah. And the very same people think the scriptures are our best guide on how to be good. All these devout people think God himself is a supremely good role model.

Here's another story, a very upsetting one, also about God testing somebody to see whether he really loved God. Imagine that, when you were a child, your father woke you one morning and said, 'It's a fine day, how would you like to come with me for a walk in the country?' You might quite fancy the idea. So off you go for a nice day together. After a while, your father stops to gather wood. He piles it up and you help him because you enjoy bonfires. But now, when the bonfire is ready to light, something terrible happens. Utterly unexpected. Your father seizes you, throws you on top of the pile of wood and ties you down so you can't move. You scream with horror. Is he going to roast you on top of the bonfire? It gets worse. Your father produces a knife, raises it above his head, and you are now in no doubt. Your father is about to run his knife through you. He's going to kill you and then set fire to your body: your own father, the father who told you bedtime stories when you were little, told you the names of flowers and birds, your dear father who gave you presents, comforted you when you were afraid of the dark. How could this be happening?

Suddenly he stops. He looks up at the sky with a strange expression on his face, as though carrying on a conversation with himself in his head. He puts away the knife, unties you and tries to explain what has happened, but you are so paralysed with horror and fear that you can scarcely hear his words. Eventually he makes you understand. It was all God's doing. God had ordered your father to kill you and offer you up as a burnt sacrifice. But it turned out to be just a tease – a test of your father's loyalty to God. Your father had to prove to God that he loved God so much that he was even prepared to kill you if God ordered him to do so. He had to prove to God that he loved God even more than he loved his own dear child. As soon as God saw that your father was really, *really* prepared to go through with it, God intervened just in time. Gotcha! April Fool! I didn't really mean it! Yes, it was a good joke, wasn't it?

Is it possible to imagine a worse trick to play on someone? A trick calculated to scar a child for life and poison a father–child relationship for ever. But that's exactly what the Bible says God did. Read the whole story in Genesis chapter 22. The father was Abraham; the child was his son Isaac.

The same story is told in the Quran (37: 99–111). Here the name of the son is not mentioned, and there is a tradition in Islam that it was Abraham's other son (with a different mother), Ishmael. In the Quran version, Abraham had a dream in which he saw himself sacrificing his son. Just a dream was enough to persuade him that Allah was telling him to do so, and he asked his son's opinion. Amazingly, the son encouraged his father to go ahead and sacrifice him. According to another Islamic tradition – this version isn't in the Quran itself – Shaytan (Satan) tried to persuade Abraham not to do this terrible deed. This would seem to make the devil the good guy in the story. But Abraham, preferring his dream, drove him away by pelting him with stones. Muslims symbolically re-enact this stoning in the annual festival called Eid.

If you were Isaac (Ishmael), could you ever forgive your father? If you were Abraham, could you ever forgive God? If anything like this happened in modern times, Abraham would be locked up for terrible cruelty to his child. Can you imagine what the judge would say if a man pleaded, 'But I was only following orders.' 'Orders from whom?' 'Well, Your Honour, I heard this voice in my head.' Or 'I had this dream.' What would you think, if you were on the jury? Would you think it was a good enough excuse? Or would you send Abraham to prison?

Fortunately there's no reason to suppose it really happened. Like most stories in the Bible, as we saw in Chapters 2 and 3, there's no good evidence for it. No evidence, indeed, that Abraham and Isaac even existed. Hardly any more than, say, Little Red Riding Hood (and that's a pretty upsetting story too, for all that everybody knows it's fiction). But the point is that, whether fiction or fact, the Bible is still held up to us as the Good Book. And its central character, God, is held up as supremely good. Many Christians still take the Bible literally as historical fact. As we shall see in Chapter 5, they think it's impossible to be good – impossible even to know what good means – without God.

In both these stories – God testing Abraham and God testing Job – I can't help feeling that the God character is not only cruel but – well – insecure. It's as though God is like a jealous wife in a novel, who is so uncertain of her husband's fidelity that she deliberately tries to trap him in unfaithfulness: persuades an attractive woman friend, perhaps, to tempt him, just to prove to herself that he'll remain loyal to her. And if God is supposed to know everything, you might think he'd know in advance how Abraham would behave when put to the test.

In the Bible, the God character often describes himself as jealous. At one point he even says his *name* is 'Jealous'! But where ordinary people are jealous of romantic rivals or business competitors, God is jealous of rival gods. Sometimes with good reason. As we saw in Chapter 1, the early Hebrews were not wholly monotheistic in the modern sense. They were loyal to Yahweh as their tribal god, but that didn't mean they doubted the existence of rival tribes' gods. They just thought their Yahweh was more powerful, and more deserving of their support. And sometimes they were tempted to worship other gods – with terrifying results if their own God caught them at it.

On one occasion, so the Bible tells us, the Israelites' legendary leader Moses was up a mountain talking with God. When Moses had been gone rather a long time the people began to wonder if he was ever coming back. They persuaded Moses's brother Aaron to collect a lot of gold from everybody, melt it down and make them a new god while Moses wasn't looking: a golden calf. They bowed down and worshipped the golden calf. That may seem odd, but worshipping statues of animals, including bulls, was quite common among local tribes at the time. Moses didn't know his people were cheating on God, but God himself could see exactly what the Israelites were up to. Mad with jealousy, he sent Moses storming down the mountain to put a stop to it. Moses seized the golden calf, burned it, ground it to powder, mixed the powder with water and made the people drink it. One of the clans of Israel, the tribe of Levi, hadn't fallen for the golden calf. So God, through Moses, ordered each Levite to pick up a sword and kill as many of the other tribesmen as they could. This amounted to a total of about three thousand dead. Even this wasn't enough to satisfy God's jealous rage. He sent a plague to ravage the people who survived. If you know what's good for you, you'd better not mess with this God character. Above all, don't you dare look at any other gods!

What had Moses been doing up the mountain with God? Among other things, he'd been taking delivery of the famous Ten Commandments, carved on tablets of stone. He carried them down with him but, such was his fury when he saw the golden calf, he dropped the tablets and broke them. Never mind: God later gave him a spare set and we are told, in two separate places in the Bible, what they said. If you ask Christians today why they think their religion is a force for good, they will very often cite the Ten Commandments. But when I've asked them what the Ten Commandments actually are, I find they often can remember only one: 'Thou shalt not kill.'

I'd say that's a pretty obvious rule for a good life. A rule that should hardly need to be carved in stone. But, as we shall see in Chapter 5, it turns out to have meant only, 'Thou shalt not kill members of thine own tribe.' God had no problem with killing foreigners. As we'll see later in this chapter, the God of the Old Testament was continually urging his chosen people to slaughter other tribes. And with a bloodthirsty ruthlessness it's hard to find in any other work of fiction. But in any case, 'Thou shalt not kill' doesn't have pride of place among the Ten

Commandments. Various traditions differ a little in how they order the commandments, but they all give prominence to Number One: 'Thou shalt have no other gods before me.' Jealous again.

> The Lord is a jealous and avenging God; the Lord takes vengeance and is wrathful. (Nahum 1: 2)

> Do not worship any other god, for the Lord, whose name is Jealous, is a jealous God. (Exodus 34: 14)

Another of God's little ways, according to the Bible, is his love of the smell of burning meat: usually non-human meat, but not always. When he ordered Abraham to truss Isaac on a bonfire, the reason as Abraham understood it was God's chronic appetite for savoury smoke. After making that last-minute intervention to save Isaac, God sent a ram to get its horns caught in a thicket nearby. Abraham got the message, killed the poor creature and gave God a fix of mutton smoke instead of Isaac smoke. The official Sunday School interpretation of the sudden appearance of the ram is that it was God's way of telling people to stop sacrificing humans and sacrifice animals instead. But the God character in the story was in the habit, in those days, of talking to people – after all, he had told Abraham to kill Isaac. So you'd think he could have simply told them in words to sacrifice sheep instead of people. Why put poor Isaac through such a terrible ordeal? You'll find, if you read the Bible, that messages are often delivered in that kind of roundabout, 'symbolic' way, rather than plainly and clearly. I can't help feeling that a really nice God would have told them not to sacrifice sheep either.

Why doesn't God seem to speak to people any more, as he did to Abraham? In parts of the Old Testament he seemingly couldn't keep his mouth shut. He seemed to speak to Moses almost every day. But nobody hears a peep from him today – or if they do, we think they need psychiatric help. Did that in itself ever make you wonder whether those old stories might not be true?

Here's another story that might make you question how nice God is. The Book of Judges, chapter 11, tells of an Israelite general called Jephthah who badly needed a victory against a rival tribe called the Ammonites. Jephthah was desperate to win, so he promised God that, if God would only give him victory over the Ammonites, he would make a burnt sacrifice of whatever or whomever he first saw on returning home after the battle. God duly gave him the victory he wanted 'with a very great slaughter'. Poor Ammonites, you might think. But it gets worse. As luck would have it, the first person who came out of the house to congratulate Jephthah was his beloved daughter. His only daughter. She came out, dancing with joy to greet her victorious father. Jephthah was horrified to remember his promise to God. But he had no choice. He had to cook his daughter. God was so looking forward to the promised smell of burning. His daughter very decently agreed to be sacrificed, asking only to be allowed to go into the mountains for two months first, 'to bewail her virginity'. After two months she did her duty and returned. Jephthah kept his promise and barbecued his daughter so God could have a nice, satisfying smoke. On this occasion God forgot the lesson of Abraham and Isaac and didn't intervene. Sorry, daughter, thank you for being so nice about it! And thanks, too, for staying a virgin, which for some reason was regarded as important for the sacrifice (verse 39).

Why was Jephthah fighting the Ammonites in the first place, and why would God have helped him gain victory? The Old Testament is filled with bloody battles. And whenever the Israelites win, the credit is given to their bloodthirsty God of Battles. The books of Joshua and Judges are largely about the campaign waged by the Israelites, after Moses had led them out of captivity in Egypt, to take over the Promised Land. This was the land of Israel, the 'land flowing with milk and honey'. God helped them take it over by exterminating the unfortunate peoples who already lived there. God's orders here were not roundabout at all, but horribly clear:

> 'When you cross the Jordan into Canaan, drive out all the inhabitants of the land before you. Destroy all their carved images and their cast idols, and demolish all their high places. Take possession of the land and settle in it, for I have given you the land to possess.' (Numbers 33: 51–3)

'For I have given you the land to possess.' What? Is that a good motive for going to war? Adolf Hitler in the Second World War justified his invasion of Poland, Russia and other lands to the east by saying that the superior German master race needed *Lebensraum*, or 'living space'. And that is exactly what God was urging his own 'chosen people' to claim by war. He was nice enough to make a distinction between those tribes who merely got in the way on the journey to the Promised Land, and those tribes who already lived in the Promised Land itself. The first group were to be offered peace. If they agreed, they got off lightly. At worst, only the men were to be killed and the women taken as sex slaves.

But less gentle treatment awaited the unfortunate peoples who actually lived in the *Lebensraum* which God had promised his chosen people:

> However, in the cities of the nations the Lord your God is giving you as an inheritance, do not leave alive anything that breathes. Completely destroy them – the Hittites, Amorites, Canaanites, Perizzites, Hivites and Jebusites – as the Lord your God has commanded you. (Deuteronomy 20: 16)

God really meant business, and his ruthless wishes were carried out to the letter. Not just during the conquest of the Promised Land but throughout the Old Testament:

> Now go, attack the Amalekites and totally destroy everything that belongs to them. Do not spare them; put to

death men and women, children and infants, cattle and sheep, camels and donkeys. (1 Samuel 15: 2)

God's orders were to kill even children. Especially boys. Girls were worth keeping for . . . well, read it for yourself and use your imagination (you won't need much).

Now kill all the boys. And kill every woman who has slept with a man, but save for yourselves every girl who has never slept with a man. (Numbers 31: 17–18)

Nowadays we'd call it ethnic cleansing and child abuse.

Theologians are embarrassed by these and the many similar passages in the Bible. They have reason to be grateful that modern archaeology and scholarship can find no evidence that any of these Old Testament stories are historically true. Theologians explain away the many horror stories as symbolic myths, moral tales like Aesop's Fables rather than history. Fair enough, although you might wonder how you could possibly find a decent moral in almost any of these terrible tales: tales of violent bloodlust, fighting for *Lebensraum*, genocidal ethnic cleansing, and treating women and girls as the property of men, to be raped and used as sex slaves.

Modern Christian theologians sometimes write off the Old Testament altogether. They point with relief to the New Testament, where Jesus comes across as a lot nicer than his terrifying heavenly father. Jesus himself was not so sure of the contrast. The gospel of John has him saying: 'I and the father are one' and 'The father is in me and I in the father' and 'Whoever has seen me has seen the father'. Nevertheless, the Jesus character in the gospels did say some pretty nice things. The Sermon on the Mount in the Book of Matthew shows Jesus as a good man, far ahead of his time. Or if he didn't exist, as a minority of scholars think, the fictional character called Jesus is a nice character. But however nice the sentiments in the Sermon on the Mount may be, the central doctrine of Christianity, as preached by St Paul, the main architect of that religion, is another matter.

The Christianity of St Paul – and that means of almost all modern Christians – regards everybody – you, me, everyone who ever lived or ever will live – as 'born in sin'. As we saw in Chapter 2, Mary's 'immaculate conception' signifies her almost unique freedom from the stain of sinful birth. Paul was obsessed with sin. You get the impression from him that God is far more interested in the sins of one species, living on one little planet, than he is in the vast expanding universe that he had created. Paul and the other early Christians believed that we all inherit the sin of Adam, the first man, who was tempted by Eve, the first woman, after she in turn was tempted by a talking snake. As we saw in Chapter 3, their sin was to eat a fruit which God had expressly forbidden them. This terrible sin – so terrible that it provoked God to drive them out of the Garden of Eden and condemn them and their descendants to a life of hard labour and pain – is thought to be inherited by all of us. According to St Augustine, one of Christianity's most revered theologians, 'Original Sin' is inherited from Adam down the male line in the semen, the fluid that carries the sperm.

Even a newborn baby, too young to have done anything, let alone anything wrong, is born with the great burden of Sin on its tiny shoulders. It's as if Paul and his Christian followers think that Sin (with a capital S) is some kind of brooding spirit: a dark, hereditary stain, rather than simply those bad things that particular people do from time to time. Born in sin, the only way we can escape everlasting damnation in the fires of hell is by being baptized and 'redeemed' by the sacrificial death of Jesus. Jesus's death was a sacrifice, like an Old Testament burnt offering, to appease God and ask him to forgive all human sin, especially the 'Original Sin' of Adam in the Garden of Eden.

Nowadays, we know that Adam never existed. Everybody who ever lived had two parents, and the line of great-great-great-grandparents goes on back through various apes and early monkeys to fish, worms and bacteria. There never was a first couple – never an Adam or Eve. There was nobody to commit the terrible sin for which we're all supposed to share the guilt. God presumably knew that, even if Paul and the early Christians didn't. And did people ever really believe in the talking snake? Actually, I'm afraid they probably did, because a disturbingly large number of people, especially in America, still do. But, setting that on one side, what about this notion of Jesus's death 'redeeming' or 'atoning for' the sins of humanity, from Adam on? It's the idea – and it really is central to the whole Christian religion – that Jesus died for our sins. He paid with his life so that our sins could be forgiven.

'Atonement' means paying for a wrongdoing. You might wonder why, if God wanted to forgive us, he didn't just forgive us. But no, that wasn't good enough for the God character. Somebody had to suffer, preferably painfully and fatally. 'Without the shedding of blood there is no forgiveness,' as the Letter to the Hebrews puts it (9: 22). St Paul often explained, in different words, that 'Christ died for our sins' (1 Corinthians 15: 3).

The idea (don't blame me, I'm just reporting the official Christian belief) is this. God wanted to forgive the sins of humankind, most prominently including the inherited sin of Adam (who never existed). But God couldn't just forgive. That would be too simple. Too obvious. Somebody had to pay for the forgiveness, in an act of sacrifice. And humanity's sin was so colossal, it couldn't just be an ordinary act of sacrifice. Nothing would do except the torture and agonizing death of God's own son Jesus. Yes, Jesus came down ('down'?) to Earth specifically so that he could be whipped and crucified, nailed to a wooden cross to die in agony and thereby pay for the sins of humanity. Nothing less than the blood sacrifice of God himself – for Jesus is regarded as God in human form – would be enough to pay for the great burden of Sin hanging round the neck of humanity.

I don't know how that strikes you, but you might well think it's a truly awful idea. At any moment leading up to the death of Jesus on the cross, an all-powerful God could have intervened – as he did in the case of Abraham's ritual sacrifice of Isaac: 'Stop, guys, it's OK. No need to hammer that nail through my beloved son's hand. I forgive you anyway. Let's all relax and celebrate the grand universal forgiveness of the sin of humanity.'

No, that seemingly obvious solution to the problem was not good enough for God. If I were writing a play about it, I might give God these lines to speak:

> Let me see, I can't just forgive them, their sin is too great. How about if I kill three thousand of them, like I did over that bit of unpleasantness with the golden calf? No, even three thousand isn't enough, not three thousand ordinary people, the sin is too great to be wiped out by killing a mere three thousand just plain folks. Tell you what, though, why don't I turn my own son into a human and have him tortured and killed on behalf of all humans? Yes, that's what I'd call a worthy sacrifice. Kill not just any old human, but God in human form! Now you're talking. That's the ticket. That would be a big enough sacrifice to atone for all the sins of humanity. Including the sin of Adam (oh, and – silly me – I keep forgetting to tell them, Adam never existed). On your way, son; sorry, but I can see no better solution. And no, you can't take the chariot of fire. I'm going to put you in

a woman's womb and you'll have to be born, brought up and educated, teenage angst and all that stuff. Otherwise you wouldn't be fully human, so I wouldn't feel you were truly representing humanity when I eventually have you crucified to save them. Don't forget, by the way, it's me myself being crucified too, because I am you and you are me.

Making fun? Yes. Savage? Maybe. Unfair? I truly don't think so, and please understand why I don't apologize. The doctrine of atonement, which Christians take very seriously indeed, is so deeply, *deeply* nasty that it deserves to be savagely ridiculed. God is supposed to be all-powerful. He created the expanding universe, galaxies hurtling away from one another. He knows the laws of science and the laws of mathematics. He invented them, after all, and he presumably even understands quantum gravity and dark matter, which is more than any scientist does. He makes the rules. The one who makes the rules has the power to forgive whomever he likes for breaking them. Yet we are asked to believe that the only way he could think of to persuade himself – *himself* – to forgive humans for their sins (most notably the sin of Adam, who never existed and therefore couldn't sin) was to have his son (who was also himself) tortured and crucified in the name of humanity. So, although the Old Testament is richer in sheer numbers of horror stories than the New, you could say that the central message of the New

Testament is a strong contender for the grim distinction of being the most horrific of all.

The disciple Judas betrayed Jesus. He led the authorities to him, and identified him with a kiss. A politician who betrays his party is called 'a Judas'. A campaign to rid the Galapagos islands of imported goats, who were ruining the natural balance, employed what were called 'Judas goats' – females marked with radio collars, who 'betrayed' the location of flocks to be exterminated. Down the ages, Judas's name has stood for the act of betrayal. But, to repeat the question we asked in Chapter 2, is this fair to Judas? God's whole plan was that Jesus had to be crucified, and so he had to be arrested. The betrayal by Judas was necessary to the plan. Why have Christians traditionally hated the name of Judas? He was only playing his part in God's plan to redeem the sins of humankind.

Even worse, the entire Jewish people has suffered persecution through the centuries because Christians have blamed them for the death of Jesus. As recently as 1938, Pius XII (a year before he became pope) spoke of the Jews as people 'whose lips curse [Christ] and whose hearts reject him even today'. Four years later, during the war (Italy was on the side of Hitler), Pope Pius spoke of Jerusalem as having the same 'rigid blindness and stubborn ingratitude' that had led it 'along the path of guilt to the murder of God'. And it wasn't just Catholics. Martin Luther, the German founder of Protestant Christianity, advocated setting fire to synagogues and Jewish schools.

Luther's pathological hatred of Jews was echoed by Adolf Hitler in 1922:

> My feeling as a Christian points me to my Lord and Saviour as a fighter. It points me to the man who once in loneliness, surrounded by a few followers, recognized these Jews for what they were and summoned men to fight against them and who, God's truth! was greatest not as a sufferer but as a fighter. In boundless love as a Christian and as a man I read through the passage which tells us how the Lord at last rose in His might and seized the scourge to drive out of the Temple the brood of vipers and adders. How terrific was His fight for the world against the Jewish poison. To-day, after two thousand years, with deepest emotion I recognize more profoundly than ever before the fact that it was for this that He had to shed His blood upon the Cross. As a Christian I have no duty to allow myself to be cheated, but I have the duty to be a fighter for truth and justice ... And if there is anything which could demonstrate that we are acting rightly it is the distress that daily grows. For as a Christian I have also a duty to my own people.

Don't take Hitler's claim to be a Christian too seriously, by the way. Whatever else Hitler was, he was a chronic liar. He may have claimed to be Christian in that speech, but in his so-called 'table talk' he was sometimes anti-Christian, although he was never an atheist and he never renounced the Roman Catholicism of his upbringing. Even if he wasn't really a sincere Christian, though, his speeches found a willing audience in a German population prepared by centuries of Catholic and Lutheran hatred of Jews. And it all started, as in the rest of Europe, with the legend that the Jews were to blame for Jesus's death.

Pontius Pilate, the Roman governor who finally approved Jesus's execution, called for water and publicly washed his hands to indicate that he took no responsibility for it. The Jews are supposed to have accepted responsibility when they cried out, 'His blood be upon us and upon our children' (Matthew 27: 25). Much of the cruel persecution suffered by Jews throughout history stems from these words. Yet – do I need to repeat the point? – the crucifixion of Jesus was the pivot of God's plan. The Jews who allegedly called for his death were only calling for what God wanted to happen anyway. By the way, don't you think that 'His blood be upon us and upon our children' sounds a rather unlikely thing for anyone to say, and suspiciously as though it was added later by a prejudiced hand?

Throughout this chapter, I've said again and again that the stories told in the Bible probably aren't true. As we saw in Chapter 2, the biblical books were written long after the events they claim to describe. If there were any eye-witnesses, most of them would have been dead by then. But that doesn't affect the main point of this chapter. Whether or not God is a fictional character, we are entitled to choose whether he's the kind of character we'd like to love and follow, as Jewish, Christian and Muslim leaders all tell us we should. What's your choice?

· 5 ·

Do we need
God in order to be good?

In the exceptionally vigorous American election campaign of 2016, the Democrat party was trying to choose between two leading candidates, Bernie Sanders and Hillary Clinton. A senior party official, Brad Marshall, wanted Hillary. He thought he'd found a way to discredit Bernie. He suspected (as though it were something wrong) that Bernie was an atheist. He wrote to two other senior party officials (Hillary herself knew nothing about it) suggesting that Bernie should be challenged, in public, to state his religion. When previously asked, he had said he was 'of Jewish heritage'. But did he really believe in God? Brad Marshall wrote:

> I think he is an atheist . . . This could make several points difference with my peeps. My Southern Baptist peeps would draw a big difference between a Jew and an atheist.

'Peeps' means 'people', and he was talking about the voters of Kentucky and West Virginia. 'Several points difference' means an important effect on votes in those two states. He thought (with good reason, unfortunately) that many Christians would rather vote for any religious person than for an atheist, even if it meant voting for someone of a different faith from themselves, in this case a Jew. Any kind of 'belief in a higher power' will do, even if it's a different higher power from their own. Opinion polls have shown the same thing again and again. There are voters who would be somewhat reluctant to vote for a Catholic, or a Muslim, or a Jew. But they still would prefer any of those to an atheist. Atheists are bottom of the list, even if the atheist is highly qualified in all other ways. Disgraceful as I think it is, it's no wonder that Brad Marshall wanted to expose the alleged atheism of the candidate he didn't favour.

The United States constitution says that 'no religious Test shall ever be required as a Qualification to any Office or public Trust under the United States'. Admittedly Marshall wasn't asking for a legal ban on atheists standing for the presidency, which really would have violated the constitution. Of course voters are allowed to notice a candidate's religion when privately casting their votes. But Marshall was deliberately appealing to voter prejudice, against the spirit of the constitution. Atheism is simply a lack of belief in anything supernatural. Like not believing in flying saucers. Or fairies. Politicians have to make decisions on things like economic policy, foreign affairs, health and social welfare, legal matters. Why should belief in the supernatural make someone take better political decisions?

I'm sorry to say that lots of people seem to think you need to believe in some sort of god, any kind of 'higher power', in order to have any chance of being moral – of being good. Or that, without belief in a higher power, you'd have no basis for knowing right from wrong, good from bad, moral from immoral. This chapter looks at the whole question of 'morals' and 'morality': what 'good' *means* as opposed to 'bad', and whether we need belief in God or gods or some sort of 'higher power' in order to be good.

So, why should somebody think you need God in order to be good? I can think of only two reasons, both bad ones. One is that the Bible, the Quran, or some other holy book tells us how to be good, and without a book of rules we wouldn't know what's right and what's wrong. We dealt with the 'Good Book' in the previous chapter and we'll return to whether we should follow it in this one. The other possible reason is that people have such a low regard for humans that they think we, politicians included, will only be good if somebody – God, if nobody else – is watching us: the theory of the Great Policeman in the Sky. Or, to update it a bit, the Great Spy Camera (or Surveillance Camera) in the Sky.

Unfortunately, there may be a certain amount of truth in that. All countries think it's necessary to have a police force. And criminals are less likely to steal or commit other crimes if they think the police are watching them. Nowadays our streets and shops are equipped with video cameras, and these often catch people doing things they shouldn't: shoplifting, for instance. Any would-be shoplifter is obviously less likely to try it on if he knows there's a camera watching him. So now imagine that a criminal believes God is watching his every move, every minute of every day. Many religious people think God even reads your thoughts and can tell in advance when you're so much as *contemplating* a bad deed. You can sort of see why those people might think a God-fearing person, including a God-fearing politician, is less likely to do bad things than an atheist. Atheists don't have to fear a great spy camera in the sky. They only – so the argument goes – have to fear real cameras and real policemen. Maybe you've heard the cynical witticism 'Conscience is knowing that someone is watching'.

The tendency to be good when you are being watched may even be quite primitive, built deeply into our brains. My colleague Professor Melissa Bateson (once an undergraduate pupil of mine at Oxford) did a remarkable experiment. In her science department at the University of Newcastle they kept an 'honesty box' to pay for the coffee, tea, milk and sugar that they used every day. Nobody was there to sell the stuff. There was a price list on the wall, and you were simply trusted to put the right amount of money in the box. It would be no surprise to learn that people are honest when somebody is looking. But what if you are alone? Would you be just as likely to put money in the box, knowing that nobody could see? I'm sure you would, but not everybody is so scrupulous, and this was what made the experiment possible.

Every week, Melissa put up the price list in the coffee room. And every week the paper was decorated with a picture at the top. Sometimes the picture was flowers: not always the same flowers, but flowers. In other weeks the picture was a pair of eyes: a different pair of eyes each time. And the fascinating result was this. In weeks when there were eyes above the price list, people were more honest. The takings in the honesty box were nearly three times as great as in the 'control' weeks when the customers had only flowers 'looking' at them. Isn't that weird? If the eyes had been a real spy camera, it would be easy to explain. But the coffee drinkers knew perfectly well that the 'eyes' were just ink on paper. Those eyes could no more see what was going on than the flowers could. It wasn't a rational calculation – 'I'd better be honest because I'm being watched.' It was irrational. Like when I stand on the top floor of a New York skyscraper and look down. I know I'm not going to fall. I'm even standing behind thick safety glass. But I still get goosebumps and a tingling of fear up my spine. It's irrational. Maybe in this case it's built into the brain by genes inherited from our ancestral past, when we needed to appreciate the danger of being high up in the trees. Perhaps you don't even need to say to yourself, 'God's eyes are watching me, so I'd better be good.' Perhaps it's an automatic, subconscious effect. Like the effect of Melissa's eyes on paper (in case you're wondering, by the way, she did the necessary sums to show the result was unlikely to be due to chance).

Whether irrational or not, it does unfortunately seem plausible that, if somebody sincerely believes God is watching his every move, he might be more likely to be good. I must say I hate that idea. I want to believe that humans are better than that. I'd like to believe I'm honest whether anyone is watching or not.

What if the fear of God is not just fear of upsetting him but of something worse – much worse? Both Christianity and Islam have traditionally taught that sinners after their death will be tormented for all eternity in hell. The Book of Revelation talks about a 'lake of fire burning with brimstone'. The Prophet Mohammed is quoted as saying that the person with the smallest punishment will have a smouldering ember placed under the bottom of his feet. 'His brains will boil because of it.' The Quran (4: 56) says of those who disbelieve its teachings, 'When their skins have been burned away, We shall replace them with new ones so that they may continue to feel the pain.' According to many preachers, you don't even have to do anything bad to be thrown into the fires of hell. It is enough to be a non-believer! Some of the greatest painters have vied with each other to produce ever more horrifying night-mare pictures of hell. The most famous work of literature in the Italian language, Dante's *Inferno*, is all about hell.

Were you threatened with hell fire as a child? Did you truly believe the threats? Were you really scared? If you can answer no to those questions, you are lucky. Unfortunately, many people go on believing the threats until they die, and it makes their lives, and especially their dying days, a misery.

I have a theory about threats of punishment. Some threats are plausible. Like, if you are found guilty of stealing you might go to prison. Other threats are very implausible. Like, if you don't believe in God, when you die you'll spend all eternity in a lake of fire. My theory is that the more plausible the threat, the less horrific it needs to be. The threat of punishment after you are dead is so far-fetched that it needs to be made really, really horrifying to compensate: a lake of fire. The threat of punishment while you are still alive is plausible (prison is a real place), so it doesn't have to involve hideous torture with your skin burning off and then being replaced to be burned off again.

What do you think of people who threaten children with eternal fire after they are dead? In this book I don't normally give my own answers to such questions. But I can't help making an exception here. I'd say those people are lucky there is no such place as hell, because I can't think of anybody who more richly deserves to go there.

Terrifying as hell is, there doesn't seem to be much clear evidence that religion makes people behave either better or worse. Some studies suggest that religious people give more generously to charity. Many give to their churches in the form of 'tithes' (meaning a tenth of their income). And churches often pass on some of that money to worthwhile charitable causes like famine relief. Or to crisis appeals after terrible disasters like earthquakes.

But a lot of the money gathered by churches goes to fund missionaries. They call it charitable giving. But is it charity in the same sense as, say, famine relief or helping people made homeless by earthquakes? Giving money for education seems a good thing to do. But if that education entirely consists of learning the Quran by heart? Or missionaries teaching children to forget their tribal heritage and learn the Bible instead?

Non-believers can also be very generous. The top three philanthropic givers in the world, Bill Gates, Warren Buffet and George Soros, are all non-believers. In 2010 a terrible earthquake devastated the already poor island of Haiti. The suffering was appalling. People around the world, whether religious or not, rallied round with offers of help and money. My own charitable foundation, the Richard Dawkins Foundation for Reason and Science, rushed to start a special charity which we called Non-Believers Giving Aid (NBGA). We recruited a dozen other non-believing, secular and sceptical organizations to join us in appealing for money from atheists, agnostics and other non-believers. Thousands of individual non-believers rallied round. Within three days, NBGA had raised $300,000. We sent every penny of it to Haiti, plus a lot more in subsequent weeks. At the same time, of course, religious charities were also gathering donations. And lots of good people went to Haiti to help. I don't tell the story of NBGA to boast that non-believers are more generous than religious believers. I actually think that, when faced with a crisis, most people all over the world are kind and generous, whether they are religious or not.

The Great Surveillance Camera in the Sky theory is sort of plausible, depressing though that is. Maybe it really does deter criminals? You might think, if so, that prison populations would have a high percentage of non-believers. Here are some figures from July 2013. They refer to the religions that convicts say they belong to, in federal prisons in the United States. Twenty-eight per cent of prisoners are Protestant Christians, 24 per cent are Catholic Christians, 5 per cent are Muslims. Most of the rest are Buddhist, Hindu, Jewish, Native American or 'unknown'. And the figure for atheists? A tiny 0.07 per cent. A convicted criminal is 750 times more likely to be Christian than atheist. Admittedly we are talking about numbers *saying* they are Christian or atheist. Who knows what figures are concealed in those 'unknowns'? More importantly, the total population of Christians in the United States is higher than the total population of atheists. But not 750 times higher. Again, the Christian figures may be somewhat inflated by the fact that prisoners can gain earlier release if they claim to be religious. It's also been suggested that the prison figures are only incidentally about religious affiliation or lack of it. Poorly educated people are more likely to end up in prison. And poorly educated people are less likely to be atheist. But, however you look at it, these figures are not promising for the Great Spy Camera in the Sky theory.

Even if the Great Spy Camera theory has some truth in it, it's certainly not a good reason to believe in the factual existence of God. The only good reason for believing anything factual is evidence. The 'Great Spy Camera' theory might be a (rather dubious?) kind of reason for hoping that *other* people will believe in God. It might bring the crime rate down. It's cheaper than installing real spy cameras or paying for more police patrols. I don't know about you, but I find that rather patronizing: 'Of course you and I are too intelligent to believe in God, but we think it would be a good idea if *other* people did!' My friend the philosopher Daniel Dennett calls it 'belief in belief': not believing in God, but believing that belief in God is a good thing. When the then Israeli Prime Minister Golda Meir was challenged to say whether she believed in God, she replied: 'I believe in the Jewish people. And the Jewish people believe in God.'

So much for the 'Great Spy Camera in the Sky' theory. I'll now turn to the other possible reason why people might think it a good idea to vote for a religious politician rather than an atheist. This is really quite different. Some people think religion is a good thing because the Bible tells us how to behave well. Without a book of rules, so the theory goes, we are adrift in a sea of uncertainty. Also, the Bible is meant to provide us with good 'role models', admired characters like God or Jesus, whom we should imitate.

But not all believers follow the Bible. Some have a completely different holy book, or no holy book at all. I'll talk here only about the Jewish/Christian Bible, because it's the only one I know well. But much the same could be said of the Quran. Do you think holy books like this are good guides to being good? Do you think the God of the Bible is a good role model? If so, you might like to take another look at Chapter 4. The Quran is even worse because Muslims are told to take it literally.

The Ten Commandments are often held up as a guide to how to live a good life. Various American states, especially in the so-called Bible Belt, are torn by fierce arguments about the Ten Commandments. On one side are Christian politicians who want to stick them up on the walls of official state buildings such as courthouses. Those on the other side usually quote the US constitution. The First Amendment to the constitution states that

> Congress shall make no law respecting an establishment of religion, or prohibiting the free exercise thereof.

That's pretty clear, wouldn't you say? The point is not that religion is forbidden. You can practise whatever religion you like, in your own way. The constitution merely forbids the establishment of an official state religion. Anybody is free to hang up the Ten Commandments privately in their own home. The constitution rightly guarantees private freedoms like that. But is it constitutional to stick them on the *public* wall of a state courthouse? Many legal experts think not.

Setting that legal question aside, let's look at the Ten Commandments themselves to see what we think about them. Are they really a valuable guide to how to be good and how not to be bad? There are two versions in the Bible, one in the Book of Exodus and one in Deuteronomy. They are pretty much the same, but different religious traditions (Jewish, Roman Catholic, Lutheran etc.) number them slightly differently. Also, Moses, in his fury about the golden calf, dropped the original stone tablets and broke them, so God later supplied him with new ones. Here's one version of the ones Moses didn't drop, as listed in Exodus chapter 20. God made a great theatrical performance of the announcement, summoning all the people to the foot of Mount Sinai and then appearing in a thunderstorm with a great trumpet blast. I've put my own comments after each commandment, and you'll probably want to add your own.

> I am the Lord your God, who brought you out of Egypt, out of the land of slavery.

For Jews that is the first commandment, although it sounds more like a statement than a commandment. For Christians it is the preamble to:

> **First Commandment:** You shall have no other gods before me.

As we saw in Chapter 4, and as God himself often said, he is a 'jealous God'.

The God character in the Old Testament was morbidly obsessed with rival gods. He hated them with a passion and was consumed by the fear that his people might be tempted to worship them. A similar obsessive loathing for rival gods persisted for centuries after the time of Jesus. After Christianity became the official religion of the Romans under Constantine, early Christian zealots rampaged around the Empire smashing what they saw as idols and we today see as priceless works of art.[*] The great statue of the goddess Athena in the ancient city of Palmyra (in modern Syria) was just one example. One of the worst offenders was the revered St Augustine. The manic determination of the early Christians to destroy images of rival gods finds its parallel today in the Muslim zealotry of ISIS and Al Qaeda.

> **Second Commandment:** You shall not make for yourself an idol in the form of anything in heaven above or on the earth beneath or in the waters below.

Again, this is all about God being jealous of rival gods. Many rival gods among neighbouring tribes were statues. The Bible drives the point home in the next verse:

> You shall not bow down to them or worship them; for I, the Lord your God, am a jealous God, punishing the children for the sin of the fathers to the third and fourth generation of those who hate me.

[*] Horrifyingly documented in Catherine Nixey's book *The Darkening Age* (London, Macmillan, 2018).

What do you think about that last sentence? God is so jealous that, if you worship a rival god, he will punish not only you but your children, your grandchildren and your great-grandchildren. Even if they were not born when you did it. Poor innocent great-grandchildren.

Third Commandment: You shall not misuse the name of the Lord your God, for the Lord will not hold anyone guiltless who misuses his name.

This means you mustn't use swear words involving God's name. Like 'God damn it!'. Or 'Don't be such a god-damn fool!' You can see why God might not like it, but it doesn't seem like a terribly serious crime, does it? Hardly worth sticking on the court-house wall. It's only 'Thou shalt not cuss', after all, and that's not the law in most countries.

Fourth Commandment: Remember the Sabbath day by keeping it holy.

God took this one very seriously indeed. In the Book of Numbers, chapter 15, the Israelites caught a man gathering sticks on the sabbath day. Gathering sticks! A pretty minor crime, you might think. But when Moses asked God what should be done about it, God was in no mood to trifle:

Then the Lord said to Moses, 'The man must die. The whole assembly must stone him outside the camp.'

Rough justice, don't you think? I don't know about you, but I think stoning is an especially horrid method of execution. It's not only painful, there's something extra nasty about the whole camp or village ganging up on one victim, like bullies in the playground. It's still done today in some Muslim countries, especially to young women caught talking to men who are not their husbands (some strict Muslims seriously think that's a crime).

Stoning no longer happens in Christian countries. One might even mischievously say that Christians are now being untrue to their holy book while the Muslim stoners are still being true to theirs. But do you think the Fourth Commandment is important enough to stick up on the court-house wall, as though it were one of the laws of the land?

The next verses justify the Fourth Commandment by pointing out that God himself took a rest on the seventh day, after his six days' labour creating the universe and everything in it.

> Six days you shall labour and do all your work, but the seventh day is a Sabbath to the Lord your God. On it you shall not do any work, neither you, nor your son or daughter, nor your manservant or maidservant, nor your animals, nor the alien within your gates. For in six days the Lord made the heavens and the earth, the sea, and all that is in them, but he rested on the seventh day. Therefore the Lord blessed the Sabbath day and made it holy.

That's typical of theological reasoning by 'analogy' – reasoning 'symbolically'. It happened this way once upon a time, so that's enough of a reason for it to happen the same way now. Actually, of course it didn't happen the first time anyway, because the universe was not created in six days, but who's counting?

> **Fifth Commandment:** Honour your father and your mother, so that you may live long in the land the Lord your God is giving you.

That's nice. It's a good thing to honour your parents. They brought you into the world, fed you, looked after you, sent you to school and many other things.

> **Sixth Commandment:** Thou shalt not kill.

This is so familiar in the language of the older King James version that I've used that here, and for the remaining commandments, rather than the more modern translation. We'd probably agree that this one is a good commandment. Perhaps that's why it's the only one that many who claim to revere the Ten Commandments can actually remember. There seems no strong objection to pinning this one up in the court-house because murder, after all, is against the law of every country. In fact, the Sixth Commandment seems almost too obvious. When Moses came down from the mountain with the stone tablets, can you imagine the people reading them and saying, 'Oh! Thou shalt not kill? Good heavens, we'd never thought of that. Fancy! Thou shalt not kill. Well, well, well. Right, I'll remember that, no more murdering people from now on.'

But although it seems obvious, the Sixth Commandment is violated in war, on a grand scale, and with the blessing of the clergy. We've already seen how, in the biblical accounts, the Israelites violated it in their fight for *Lebensraum* against the unfortunate peoples who already lived in the Promised Land – and did so on explicit orders from God. In the First World War, British soldiers were ordered to kill German soldiers. And German soldiers were given similar orders to kill their enemies. Both sides thought God was egging them on, which inspired the poet J. C. Squire to write:

> God heard the embattled nations sing and shout
> '*Gott strafe England*' and 'God save the King!'
> God this, God that, and God the other thing –
> 'Good God!' said God, 'I've got my work cut out!'

Orders to kill, with God's apparent blessing, have been given to soldiers in wars throughout history.

Think about this. In those American states that execute murderers, the accused is brought to trial: this can last weeks or months, and a prosecuting lawyer has to convince a jury of guilt 'beyond reasonable doubt'. Numerous appeals can be lodged before the death penalty is actually carried out. Finally, a solemn death warrant has to be signed by the state governor, who usually takes the responsibility very seriously. And then, on the morning of the execution, there is a grisly ritual of a last, favourite breakfast. But when a British soldier kills a German soldier in war, the German soldier has – as far as the British soldier knows – committed no crime. He has not been tried in court. He has not been formally sentenced to death, he cannot call a lawyer and he has no right of appeal. He may not even have volunteered for the army but simply been called up, against his will. And then we're ordered to shoot him. In the Second World War, bomber crews on both sides were ordered to kill thousands of civilians, again without trial. Thou shalt not kill?

In Britain you could gain exemption from military service by declaring that you were a conscientious objector who refused to kill – but then you had to go before a tribunal to justify your objection to killing, and it was quite difficult to convince them. The easy way to be allowed not to fight was to have parents who belonged to a pacifist religion such as the Quakers. But if you'd thought it out for yourself, perhaps even written a PhD thesis on the immorality of war, you still had to convince the tribunal you should be allowed to stay out of the army. If you succeeded, you could drive an ambulance instead. I probably would have failed to convince them. But I would have secretly shot to miss.

What the Sixth Commandment originally meant was 'Thou shalt not kill members of thine own tribe.' (Unless, of course, they gather sticks on the sabbath or commit other unforgivable crimes!) We know that, because the God character ordered his people to kill other tribes with abandon and with relish.

Seventh Commandment: Thou shalt not commit adultery.

That sounds straightforward enough. Don't have sex with somebody if either of you is married to somebody else. But perhaps you can imagine circumstances where it should be relaxed. Like when somebody in an unhappy and long broken marriage falls deeply in love with somebody else. As we'll see later, some people think moral rules are absolute and unbreakable under any circumstances. Other people think rules should be relaxed depending on the particular case. Anyway, many people would say that each individual's love life is a private matter and not a matter for a commandment stuck up in a state court-house as though it were a law of the land.

Eighth Commandment: Thou shalt not steal.

As with 'Thou shalt not kill', there seems no objection to putting this up in the court-house. Stealing, like murder, is against the law in all countries anyway.

Ninth Commandment: Thou shalt not bear false witness against thy neighbour.

Yes indeed. Don't bear false witness against – that is, tell lies about – anybody, neighbour or not. Again, it's a cornerstone of the law that witnesses, especially when under oath, must 'tell the truth, the whole truth and nothing but the truth'.

> **Tenth Commandment:** Thou shalt not covet thy neighbour's house, thou shalt not covet thy neighbour's wife, nor his manservant, nor his maidservant, nor his ox, nor his ass, nor any thing that is thy neighbour's.

'Covet' is a somewhat outdated word for 'envy', with the added element of seeking to possess the envied thing or person. It can be hard not to envy somebody who is much more fortunate than you are. But it's surely not a matter for the law, so long as you don't actually go out and grab the thing you covet. Even that, according to some political revolutionaries, might be justified. They think the state is justified in seizing private wealth and using it for everybody. I'm not a communist or an anarchist, but perhaps you can see where they are coming from? Other people, who call themselves libertarians, go to the opposite extreme. They think that even taxation is a form of theft, robbing the rich to pay the poor. The legendary bowman Robin Hood did exactly that, and he has a certain romantic appeal in some quarters. Like his more modern equivalents, the Wild West's Jesse James and the Irish highwayman Willie Brennan.

By the way, notice that the Tenth Commandment counts the neighbour's wife, and his servants, among his possessions, like his house or his ox. What do you think of the idea that a woman is the property of some man: one of his possessions, a 'thing' that he owns? I think it's a horrible idea, but it has long been deeply embedded in many cultures and we still see it today in places such as Pakistan and Saudi Arabia, where it is sanctioned by the state religion. Some people (not I) think that's a good enough reason to 'respect' it. You may have heard the phrase 'It's part of their culture', with the implication that we have to respect it. Saudi Arabia, as I write, has only just passed a law allowing women to drive. A married woman is still not allowed to open a bank account without her husband's permission. She is not allowed out of the house unless accompanied by her husband or by a male relative – who can be a tiny male child. Just picture the scene: a grown woman, university-educated perhaps, has to ask her eight-year-old son for permission to leave the house. And he has to come with her to serve as her male 'protector'. Those woman-hating laws are inspired by Islam.

I can imagine that, if the Tenth Commandment were pinned up in an American court-house, plenty of women would have something to say about it. At very least we might add, in the interests of equality (and moving with the times), 'Thou shalt not covet thy neighbour's husband. Nor her Jaguar. Nor her doctoral degree.'

Well, of course, the Ten Commandments are out of date. It's unfair to blame the Bible for being written thousands of years ago when men owned their wives, and their most prized possessions were their slaves. Of course we've moved on since those bad old days. But isn't that the whole point? Yes, we have moved on. And that's precisely why we shouldn't be getting our morals, our 'right and wrong', our 'do and don't' from the Bible. And as a matter of fact we don't get them from the Bible. If we did, we'd still be stoning people to death for working on the sabbath. Or for worshipping the wrong gods.

'But', some may say, 'that's just the Old Testament. Let's get our morals from the New Testament instead.' Well, yes, that might be a better idea. Jesus said some pretty nice things, in the Sermon on the Mount, for instance. Certainly very different from anything in the Old Testament. But how do we know which statements in the Bible are good, which bad? How do we decide? That decision has to be based on something outside the Bible: otherwise it's circular reasoning, unless you invent a rule like 'later verses supersede earlier ones'. By the way, Islam does have exactly that rule, but unfortunately it ends up going the wrong way. The Prophet Mohammed said some quite nice things during his earlier time in Mecca. But later, after he moved to Medina, he became, for reasons to do with the historical circumstances, much more warlike. Many of the awful things done in the name of Islam can be justified using later 'Medina verses' in the Quran, which contradict – and supersede, according to the official doctrine – earlier, nicer, 'Mecca verses'.

Back to the Christian Bible. There's nothing in it that says, 'Forget about the Old Testament, just read the New Testament to find out what's right or wrong.' Jesus could have said that. In fact (Matthew 5: 17), he said exactly the opposite:

> Do not think that I have come to abolish the Law or the Prophets; I have not come to abolish them but to fulfil them. I tell you the truth, until heaven and earth disappear, not the smallest letter, not the least stroke of a pen, will by any means disappear from the Law until everything is accomplished.

Also in Luke (16: 17):

> It is easier for heaven and earth to disappear than for the least stroke of a pen to drop out of the Law.

'The Law', to a Jew like Jesus, meant certain books of the Old Testament. Jesus seems to have taken a rather rosy view of the Old Testament. In Matthew 7: 12 he states the rather nice principle which we know as the Golden Rule (treat other people the way you'd like them to treat you) and goes on to say it is the central message of the Old Testament:

> So in everything, do to others what you would have them do to you, for this sums up the Law and the Prophets.

It's true you can find something that sounds a little bit like the Golden Rule in the Old Testament (and you can find older, more precise versions of the Golden Rule in texts from ancient Egypt, India, China and Greece):

> Do not seek revenge or bear a grudge against one of your people, but love your neighbour as yourself. I am the Lord. (Leviticus 19: 18)

But it's a great exaggeration to say it's the main message of the Old Testament. As we saw in Chapter 4, God was himself pretty expert at bearing grudges. And there's any number of verses of the Old Testament which preach vengeance.

> If anyone injures his neighbour, whatever he has done must be done to him: fracture for fracture, eye for eye, tooth for tooth. As he has injured the other, so he is to be injured. (Leviticus 24: 19)

By the way, that's another thing that comes straight from Babylon, in this case the 'Code of Hammurabi'. Hammurabi was a renowned Babylonian king, and his rulebook was written down about a thousand years before the Old Testament.

Here's another version from the Bible, the Book of Deuteronomy:

> Show no pity: life for life, eye for eye, tooth for tooth, hand for hand, foot for foot. (Deuteronomy 19: 21)

I suppose you could say it's a kind of negative version of the Golden Rule. But it doesn't sound so nice the negative way round, does it? Jesus himself (Matthew 5: 38) went out of his way to say the opposite, even quoting that very verse from the Old Testament:

You have heard that it was said, 'Eye for eye, and tooth for tooth.' But I tell you, Do not resist an evil person. If someone strikes you on the right cheek, turn to him the other also. And if someone wants to sue you and take your tunic, let him have your cloak as well. If someone forces you to go one mile, go with him two miles.

I don't think there's ever been a clearer or more generous repudiation of the idea of vengeance. It puts Jesus far ahead of his time. And far ahead of the Old Testament God.

Yet Jesus himself was not above vengeance. Even discounting the stories in the infant gospel of Thomas, the canonical gospels of both Matthew and Mark tell how he took petty revenge on, of all things, a fig tree:

Early in the morning, as he was on his way back to the city, he was hungry. Seeing a fig tree by the road, he went up to it but found nothing on it except leaves. Then he said to it, 'May you never bear fruit again!' Immediately the tree withered. (Matthew 21: 18)

Mark's version (11: 13) adds that the reason there were no figs on the tree was that it was too early in the year. Poor fig tree; it was simply not yet the fruiting season.

Christians are understandably embarrassed by the story of the fig tree. Some say it never happened, like the stories in the infant gospel of Thomas. Others just ignore it and concentrate on the nice bits of the New Testament. Yet others say it was 'symbolic'. There never was an actual fig tree. It was some kind of metaphor for the nation of

Israel. That's a favourite dodge of theologians, had you noticed? If you don't like something in the Bible, say it's only symbolic, it never really happened, it's a metaphor to convey a message. And of course they get to choose which verses are metaphors and which are to be taken literally.

There are other places in the official gospels where Jesus comes across with some of the nastiness of his Old Testament 'father'. In Luke 19: 27 he says of people who don't want him to reign over them as their king, bring them 'hither and slay them before me'. Rather surprisingly in view of Roman Catholic worship of his mother Mary, Jesus himself was not very nice to her. On the occasion of his first miracle, turning water into wine at a wedding feast, when his mother approached him, Jesus said, 'Woman, what have I to do with thee?' Perhaps it sounded less cruel in the original Aramaic than when translated into the English of the King James version. One of the modern translations, the New International version, sticks 'Dear' in front of 'woman', which at least changes the tone for the better. (A classical scholar friend tells me the Greek word used here for 'woman' can sometimes have a sort of 'dear' meaning.) And, to be fair, since the whole story of turning water into wine certainly isn't true, there's a good chance Jesus's apparent put-down of Mary at the wedding didn't really happen either.

Whether it happened or not, this is not the only story where Jesus comes across as a surprise choice of role model for family values:

'If anyone comes to me and does not hate his father and mother, his wife and children, his brothers and sisters – yes, even his own life – he cannot be my disciple.' (Luke 14: 26)

On another occasion, Jesus was speaking to a crowd of people and was told that his mother and his brothers were waiting, hoping to have a word with him. Again, a put-down:

Someone told him, 'Your mother and brothers are standing outside, wanting to speak to you.' He replied to him, 'Who is my mother, and who are my brothers?' And he waved his hand towards his disciples and said, 'Here are my mother and my brothers.' (Matthew 12: 48)

At other times, Jesus comes across as not so much bad as ignorant, and in a not very nice way. In the Gadarene region, Jesus came upon a pair of men 'possessed' by 'demons' (Matthew 8). 'They were so violent that no one could pass that way.' Probably schizophrenia, then, or some other mental illness, but Jesus followed the false belief of his time, the belief in 'demons'. He commanded the demons to come out of the men. But now the demons had nowhere to go, so he told them, instead, to enter a herd of pigs that were feeding nearby. The demons did, and the poor pigs (now known proverbially as the Gadarene Swine) stampeded headlong over a steep cliff and drowned. Not a nice story. Of course I wouldn't normally blame a man of the first century for ignorance of mental illness. Judging people of an earlier time by the standards of your own time is one of the things good historians just don't do. But Jesus was supposed to be no ordinary man. He was supposed to be God. Shouldn't God have known better?

Jesus was not a bad man, simply a man of his time. Imagine how impressive it would be if Jesus had said, 'Verily I say unto you, there are no demons, nothing that could fly out of a man and into pigs. This man has an affliction in his head. There are no demons anywhere.' Even better, imagine how impressed we'd be if Jesus had told his disciples that the Earth orbits the sun, that all living creatures are cousins, that the Earth is billions of years old, that the map of the world changes over millions of years . . . But no, his wisdom, impressive though it was in many ways, was the wisdom of a good man of his time, not a god. Just a man, though a good one.

And imagine how impressed we'd be if the Prophet Mohammed, channelling God, had said, 'O Believers, the Sun is a star like any of the other stars in the sky. Just much nearer than them. It seems to rise in the east and travel across the sky till it sets in the west. But truly it is just the Earth spinning which makes it look that way.' Alas, no, what he actually said was, 'The Sun sets in a marsh.'

Or suppose Elijah, or Isaiah, had said, 'Hear, O Israel, the word of the Lord your God. The Lord has revealed to me in a dream that nothing can travel faster than light.' Instead, all we get from them is orders to worship only one God, plus lots of other rules for how to live – all things that might occur to men of their own time.

You can find some nice verses in the Bible, some even in the Old Testament – though not many, in my experience. But how do we *decide* which verses to ignore because they are nasty, which verses to promote because they are nice? The answer has to be that we have some other criterion for deciding, some method of judging what's nice and what's nasty. A reason that doesn't come from the Bible itself. But then, whatever that criterion turns out to be, why don't we just use it directly? If we have some independent criterion for deciding which biblical verses are good and which bad, why bother with the Bible at all?

But, you may say, it's all very well talking about an independent set of standards. It does seem to be there, but what is it? How do we, as a matter of fact, decide what's good and what's bad (and therefore, incidentally, which verses of holy books are nice and which nasty)? That is the subject of the next chapter.

· 6 ·

How do we decide what is good?

Like all other animals, we humans are the product of hundreds of millions of years of evolution. Brains evolve like all other parts of the body. And that means that what we do, what we like doing, what feels right or wrong, also evolve. We inherit from our ancestors a liking for sweet things and a 'yuck' reaction to the smell of decay. We inherit evolved sexual desires. All those are easy to understand. In moderation, sugar is good for us, although too much is not. We now live in a world where too much sugar is readily available. But this was not true of our wild ancestors on the African savanna. Fruit was good for them, and many fruits contain moderate amounts of sugar. It was impossible to get too much sugar, so we evolved an open-ended appetite for it. The smell of decay is associated with dangerous bacteria. It benefited our ancestors to avoid decaying meat, and that included a revulsion to the smell. It's obvious why we evolved a desire for the opposite sex. Sexual desire leads to babies, and those babies then carry the genes that give them sexual desires when they grow up. We are all descended from an unbroken line of ancestors who mated with a member of the opposite sex, and we have inherited their desire to do so.

But now for something less easy to understand. We seem also to have inherited a desire to be nice to other people. To be friends with them, spend time with them, cooperate with them, sympathize when they are in distress, help them when they are down. The evolutionary reason to be nice is hard to explain, and must wait until Chapter 11, after the chapters on evolution itself. Meanwhile I can only ask you to accept that niceness, of a special limited kind, is part of our evolutionary heritage, like sexual desire. And it probably feeds into our sense of right and wrong. We have evolved moral values, inherited from our remote ancestors.

And yet that can only be a part of the answer to the question that heads this chapter. This must be so, if only because our view of right and wrong changes as the centuries go by, and changes on a historical time-scale much too fast to represent evolutionary change.

You can see it as the decades go by. It's almost like 'something in the air'. Of course it isn't literally in the air. It's a combination of lots of things, so it sort of feels like 'in the air' because it can't be pinned down to any one place. The dominant moral values of the twenty-first century, in which we are now living, are noticeably different from those of even a hundred years ago. They're even more different from those that prevailed in the eighteenth century. Then, keeping slaves was simply what people did – including my ancestors in Jamaica, I'm sorry to say – and they thought civilization would collapse if the slaves were freed. The great Thomas Jefferson, third President of the United States and the main author of the US constitution, kept slaves. So did George Washington, the first President. Let's at least hope they (and my ancestors) didn't know the appalling conditions in the ships that transported the slaves from West Africa.

By the way, it wasn't just white Europeans and Americans who took slaves from Africa. While Europeans were taking slaves from West Africa, Arabs were taking them from East Africa. Swahili, which has become the dominant language of equatorial East Africa, developed as the language of the Arab slave trade. It contains many words of Arabic origin. African chiefs also kept slaves themselves, as well as capturing and selling them to European and Arab traders. Not surprisingly, since the Bible's morality was of its time, slavery is not condemned there. Even the New Testament is full of exhortations like:

> Slaves, obey your earthly masters with respect and fear, and with sincerity of heart, just as you would obey Christ. Obey them not only to win their favour when their eye is on you, but like slaves of Christ, doing the will of God from your heart. (Ephesians 6: 5)

Here's another:

> All who are under the yoke of slavery should consider their masters worthy of full respect, so that God's name and our teaching may not be slandered. (1 Timothy 6)

The revulsion against slavery which we feel today is just one example of a change 'in the air'. Abraham Lincoln, another of America's most revered presidents, was Charles Darwin's exact contemporary, born the same February day in 1809. Darwin was passionately against slavery, and Lincoln actually freed the slaves in America. Yet it would not have occurred to either Darwin or Lincoln that Africans could be the equal of what they called 'the civilized races'. Darwin's friend Thomas Henry Huxley was an even more obviously advanced, liberal thinker. Yet in 1871 he wrote this:

> No rational man, cognizant of the facts, believes that the average negro is the equal, still less the superior, of the white man. And if this be true, it is simply incredible that, when all his disabilities are removed, and our prognathous relative has a fair field and no favour, as well as no oppressor, he will be able to compete successfully with his bigger-brained and smaller-jawed rival, in a contest which is to be carried on by thoughts and not by bites. The highest places in the hierarchy of civilization will assuredly not be within the reach of our dusky cousins.

And President Lincoln said this, in 1858:

> I will say, then, that I am not, nor ever have been, in favor of bringing about in any way the social and political equality of the white and black races; that I am not, nor ever have been, in favor of making voters or jurors of negroes, nor of qualifying them to hold office, nor to

intermarry with white people; and I will say, in addition to this, that there is a physical difference between
the white and black races which I believe will forever
forbid the two races living together on terms of social
and political equality. And in as much as they cannot
so live, while they do remain together there must be the
position of superior and inferior, and I as much as any
other man am in favor of having the superior position
assigned to the white race.

Truly, whatever was 'in the air' in the nineteenth
century, something very different hovers about us today.
It's a poor historian who would condemn Lincoln and
Darwin and Huxley as racists. They were as near to being
non-racist as men of their time ever got. They were men of
the nineteenth century. If they'd been born two centuries
later they'd have been horrified by those two quotations.

You don't even have to wait one century to notice a
change in moral values. In Chapter 5 we considered the
bomber crews who, on both sides, slaughtered masses
of civilians in the Second World War. To start with, the
bombing focused on industrial centres like Coventry in
Britain and Essen in Germany, where arms were being
manufactured. Bombing was inaccurate in those days,
and civilian casualties were inevitable. But both sides
resented their civilian deaths. They retaliated. And later
in the war, the bombing raids were scaled up: civilian
casualties ceased to be a byproduct and became the
objective. Between 13 and 15 February 1945, 722 British

and 527 US planes flattened the ancient and beautiful German city of Dresden with high explosives and fire-bombs. The exact number of civilian casualties will never be known, but realistic estimates put it at more than 100,000. That's comparable to the figures for each of the atomic bombs that destroyed Hiroshima and Nagasaki in August 1945.

Now move on half a century. Unhappily there are still wars, but they're nowhere near so terrible as the two world wars. In the two Gulf Wars, although there were still civilian casualties, these were treated as unfortunate mistakes. Politicians apologized for them and explained that they were 'collateral damage', byproducts of attacks on 'legitimate' military targets. It's partly that electronic technology has advanced. Guided missiles, with satellite control and other navigation systems, can cruise accurately to a particular address keyed into their onboard computer. Very different from the indiscriminate bombing of Dresden, London and Coventry. But the moral climate 'in the air' has moved on too. In the Second World War, people like Hitler, and Marshal of the Royal Air Force Sir Arthur 'Bomber' Harris, positively wanted to kill civilians. The modern equivalents of Bomber Harris (his less complimentary nickname in the RAF was 'Butcher Harris') go out of their way to apologize when a civilian is killed by a stray missile.

Can you believe how recently women were first allowed to vote? In Britain, women gained the same voting rights as men in 1928. Up to 1918 no women could vote, and then only those who had reached the age of 30 and met certain property and/or educational criteria. At that time men could vote at 21. The United States gave women the vote in 1920 (finally catching up with various individual states within the Union). French women couldn't vote till 1945. And Swiss women not till 1971. As for Saudi Arabia, don't even ask! The point is that something changes, something spreads 'in the air' such that, as the decades go by, the things that people find acceptable change. Dramatically quickly. Before women had the vote in Britain, nice, decent men could be heard saying things like, 'Women are sweet and pretty and all that, but they can't think logically. They certainly shouldn't be allowed to vote.' Can you imagine anyone saying that nowadays?

My friend the psychologist Steven Pinker has written a great (in both senses) book called *The Better Angels of our Nature* (the title is a quote from Abraham Lincoln). He shows how, over the centuries, over the millennia, we humans have been getting nicer, gentler, less violent, less cruel. The change has nothing to do with genetic evolution and nothing to do with religion. Whatever is 'in the air' has been moving in what we can broadly see as the same direction from century to century.

It's the same direction, but is it the 'right' direction? Well, I think so and I expect you do too. Is that only because we are twenty-first-century people? I'll leave that for you to decide. But when, in Chapter 4, we judged the God character in the Old Testament, we were judging him by the standards of our own century. Just as a good historian doesn't look down on Abe Lincoln for his racial prejudices, so the historian might hesitate to think the worse of the God character for the truly terrible things he did. To Isaac at the hands of his father, for instance. And to Jephthah's daughter. And to the poor Amalekites and the other tribes whose 'land of milk and honey' the Israelites were told to covet. The God character, in the books of the Old Testament, was only acting out the moral values that were 'in the air' at the time. But, although we may make allowances for his moral values (or rather the moral values of the Jews in Babylon who wrote the Old Testament), that doesn't stop us resolving firmly to do things differently in our time. And we are entitled to oppose those fundamentalists of today who try to drag us back to those times.

Right then, moral values are 'in the air' and they change from century to century, even decade to decade. But, in addition to our evolutionary past, where do they actually come from? And why do they change? Partly the changes come from ordinary conversations, in cafés and in pubs and around dinner tables. We learn from each other. We hear stories about people we admire, and vow to imitate them. We read novels, or opinion pieces in newspapers, listen to podcasts or speeches on YouTube, and change our minds. Parliaments and congresses debate questions and change the law, step by step. Judges interpret the law in ways that change as the decades go by.

Before 1967, British men could go to prison for homosexual acts in private. Now, after decades of work against persistent prejudice, being gay has become normal, and gay people can claim the same respect as anyone else. It was parliamentary votes (after a long, hard struggle by the suffrage campaigners) that gave women the vote, in country after country, during the course of the twentieth century. And we may be sure members of parliament and congresses were influenced by letters they received from their constituencies and congressional districts. Decisions in courts, by judges and juries, also serve to move the climate of opinion along, as the decades go by. And we mustn't forget academic books, and lectures in universities. Scholars who make a study of moral values, of right and wrong – moral philosophers – have an influence on the changes 'in the air'. I'll say a bit about moral philosophy here to round off this chapter.

There are various schools of moral philosophy. I shall talk about only two of them: absolutists and consequentialists. They take very different views of how to make moral judgements. Absolutists think some things just are right and some things just are wrong. No argument. Rightness or wrongness is just a fact, just plain true, like the statement in geometry that parallel lines never meet. An absolutist might say, 'Killing another human being is just plain wrong. Always is, always has been, always will be.' An absolutist of that type might say abortion is murder because an embryo is a human being. Some absolutists would even apply that argument to a fertilized egg, a single cell.

Consequentialists judge right and wrong differently. You'll have guessed from the name that they care about the *consequences* of an action. For example, who *suffers* as a consequence of an abortion? Or who suffers as a consequence of refusing an abortion? Let's imagine a conversation between a consequentialist (Connie) and an absolutist (Abby). It gives an idea of how moral philosophers think, and argue. Philosophers, from Plato through Hume to this day, are fond of making up dialogues between imagined arguers, and I am following their example. Notice, as we go, how quickly philosophers move from reality to 'thought experiments'.

Abby: Thou shalt not kill another human being. A fertilized egg is a human being. Therefore abortion, even of a single fertilized egg cell, is murder. I've heard a woman friend say, 'A woman has an absolute right to do what she wants with her own body. That includes the right to kill an embryo which is in her body. It's nobody else's business but hers.' But the embryo is another human being. It has rights too, even though it is inside her body.

Connie: Your woman friend's argument is an absolutist argument, like yours. She claims an 'absolute right' to her own body and everything inside it. That's absolutism, although a different kind of absolutism from yours. And you and she come to opposite conclusions.

But I'm a consequentialist. I ask who suffers. You can define a fertilized egg as a human being if you like. But it doesn't have a nervous system, so it can't suffer. It doesn't know it's been aborted, feels no fear or regret. A woman has a nervous system. She can suffer if she's made to have a baby that she doesn't want and can't afford to look after. You and your woman friend are both absolutists. She's a 'women's rights absolutist'. You are (I suspect) a religious absolutist. I agree with her conclusion, but for a different reason. Her reason is absolutist: a woman's absolute right to control what happens in her own body. My reason is consequentialist. An embryo can't suffer but a woman can.

Abby: Well, I agree that a single-cell embryo can't suffer, but it has the *potential* to become a fully fledged human being. The abortion is depriving it of that opportunity. Wouldn't you call that a 'consequence'? Perhaps I'm a kind of consequentialist, too? More so than my woman friend, anyway!

Connie: Yes, I agree that depriving the embryo of future life is a consequence. But since the cell doesn't know about it, and feels no pain or regret, why worry? Also, every time you refuse to have sex you are potentially depriving a future human being of the opportunity for a life. Had you thought of that?

Abby: At first sight, that's not a bad point. But still, before the sperm meets the egg there's no particular person there. By avoiding sex, you're not depriving an individual person of existence, because there are millions of sperms and millions of potential individuals. Once a sperm is inside an egg, a particular individual person has begun. No other person. Before that moment there could be a million lives, so you can't say you are depriving any one person of existence.

Connie: But if you talk about a fertilized egg as 'a particular individual person', you're implying an indivisible entity. Do you know any identical twins? They start off as one fertilized egg. Then later they split and become two individuals. Next time you meet a pair of identical twins, why not ask them which one is the 'person', which one the zombie.

Abby: Hm, OK, I see what you mean. That's an alarmingly good point. Perhaps I'd better change the subject. If all you care about is who suffers as a consequence of your actions, what's wrong with cannibalism? I'm sure you wouldn't kill anybody to eat them, but how about eating somebody who's already dead and can't suffer?

Connie: His friends and relatives would hate it. That's a consequence! An important one. People's feelings matter. But only those with nervous systems have feelings. A pregnant woman who desperately doesn't want another baby has feelings. The embryo inside her doesn't.

Abby: Sticking to my cannibalism example, suppose the dead person has no friends or relations. Nobody would suffer as a result of your eating him.

Connie: Well, now we come to what we call the 'slippery slope' argument. You might feel safe at the top of a precipitous hill, but if the slope down the hill is slippery and you set one foot on it, before you know what's happened you find yourself sliding down to the bottom, where you don't want to be. You are right that nobody would suffer if I eat an already dead person who has no friends or relations to care. That's the top of the slippery slope. But our society has a deep and well-established taboo against cannibalism. We are revolted by the very idea. If once we break through the taboo, we are in danger of sliding down the slippery slope. Who knows where it will end? The taboo against cannibalism is useful, like a safety railing at the top of a dangerously steep slope.

Abby: Well, I can apply the slippery slope argument to abortion, too. I agree that an early embryo can't feel pain or fear or sorrow at being aborted. But there's a slippery slope all the way to the moment of birth and beyond. If you allow abortion, isn't there a risk of sliding down the slippery slope all the way past the moment of birth? Mightn't we end up murdering one-year-old babies just because they are a nuisance? Then two-year-olds. And so on?

Connie: Yes. I must say that sounds at first like a fair point. But the moment of birth is a pretty good barrier – a pretty good 'safety railing' – one that we are accustomed to respecting. Although it hasn't always been so. In ancient Greece they would wait till a baby was born, take one look at it and then decide if they wanted to keep it. If not, they'd leave it out on a cold hillside to die. I'm so glad we don't do that now. By the way, late abortions are very rare, and only done for urgent reasons, usually to save the mother's life. The vast majority of abortions are early. And did you realize that many conceptions abort spontaneously without the woman even knowing she was pregnant?

But actually, although I just used the slippery slope argument, I must admit that I prefer to do away with barriers and lines altogether. You absolutists want to draw a hard and fast line between human and non-human. Does an embryo become human at the moment of conception, when the sperm first joins the egg? Or at the moment of birth? Or at some point between, in which case precisely when? I prefer to ask a different question. Not 'When does it become human?' but 'When does it become capable of feeling pain and emotion?' And there is no sudden moment when that happens. It's gradual.

The same is true in evolutionary time. We don't kill humans to eat them. We do kill pigs to eat them. Yet we are cousins to pigs, which means that, if we follow our ancestors backwards and pigs' ancestors backwards, sooner or later we'll hit the shared ancestor. Think back through our family tree. On the way to the ancestor we share with pigs, we'll pass through ape men, monkey-like creatures and so on. Now, imagine that those ape-man species had not gone extinct. At what point would you say, 'Right, that's it, from now on back they aren't human any more'? You are an absolutist who wants to draw an absolute line between humans and animals. But I'm a consequentialist who prefers not to draw lines at all, if we can avoid it. In this case my question would not be 'Is this creature human?' but 'Can this creature suffer?' And I presume some animals can suffer more than others. Including pigs, by the way.

Abby: Your moral arguments seem logical. But even you have to start with some kind of absolutist belief. In your case you start by simply saying 'Causing suffering is wrong.' You offer no justification for that.

Connie: Yes, I admit that. But I still think my absolutist belief that 'Causing suffering is wrong' makes more sense than your absolutist belief, 'It says so in my holy book.' I think if anybody were to torture you you'd pretty quickly agree.

You can carry on the argument between Abby and Connie yourself. I hope I've taken it far enough to show you the kind of way moral philosophers argue. You've probably guessed that absolutists are often religious, although it's not a hard and fast rule. The Ten Commandments are clearly absolutist. So, usually, is the very idea of living by a set of rules.

It's possible for non-religious philosophers to devise rule-based moralities, however. Various schools of moral philosophers, called deontologists, believe you can justify rules on grounds other than simply looking up statements in a holy book. For example, the great German philosopher Immanuel Kant stated a rule called the Categorical Imperative: 'Act only according to that maxim whereby you can, at the same time, will that it should become a universal law.' The key word here is 'universal'. A rule encouraging stealing is ruled out, for example, because if it were universally adopted, that is, if everybody stole, no one would benefit: thieves prosper only in a society dominated by honest victims. If everybody told lies all the time, lying would cease to have meaning because there wouldn't be any reliable truth to compare it with. A modern deontological theory proposes that we should devise our moral rules behind a 'veil of ignorance'. Pretend you don't know whether you are rich or poor, gifted or untalented, beautiful or ugly. Those facts lie concealed behind the imagined 'veil of ignorance'. Now devise the system of values you'd like to live under, given that you can't know whether you will be at the top of the heap or the bottom. Deontology is interesting, but I'll say no more about it here in a book about religion.

The argument about when, in the womb, a 'person' begins is very much a religious argument. Many religious traditions see the immortal soul as entering the body at some definite moment. Roman Catholics think it's the moment of conception. The Catholic Doctrine of the Faith entitled *Donum Vitae* is very clear on the point:

> From the time that the ovum is fertilized, a new life is begun which is neither that of the father nor of the mother; it is rather the life of a new human being with his own growth. It would never be made human if it were not human already . . . Right from fertilization is begun the adventure of a human life.

It would seem that whoever wrote that had never thought of the 'identical twin' argument: the one Connie the consequentialist used.

You've probably guessed that my sympathies lie with Connie more than with Abby. I must admit, however, that consequentialist thought experiments sometimes lead in uncomfortable directions. Suppose a coal miner is trapped underground by a fall of rock. We could rescue him, but it would cost a lot of money. What else might we do with that money? We could save a lot more lives and reduce a lot more suffering by spending it on food for starving children around the world. Shouldn't a true consequentialist abandon the poor miner to his fate, never mind his weeping wife and children? Maybe, but I wouldn't. I couldn't bear to leave him underground. Could you? But it's hard to justify the decision to rescue him on purely consequentialist grounds. Not impossible but hard.

Let's return to the main topic of this chapter. Do we need God in order to be good? I've spent quite a lot of time on moral philosophy, but moral philosophy is just one of the routes through which moral values change. Along with journalism, dinner-table conversations, debates in parliamentary chambers and student unions, legal judgments and so on, moral philosophy contributes to the shifting 'something in the air' which makes twenty-first-century morality different from, say, eighteenth-century morality, according to which slavery was a good thing. By the way, there seems no obvious reason for the trend to stop. What will twenty-second-century morality look like?

Our modern morality, whether we are religious or not, is very different from biblical morality. Or Quranic morality. Thank goodness. And the Great Spy Camera in the Sky is surely not a praiseworthy reason to be good. So perhaps we should all give up the idea that we 'need God in order to be good'.

Would that mean we should all give up believing in God? No. Not for that reason alone. He might still exist even if we don't need him in order to be good. A god could be bad by our own moral standards, like the God character we met in Chapter 4, and that still wouldn't mean he can't exist. Evidence is the only reason to believe in the existence of anything. Is there any evidence, any good evidence anywhere, for any kind of god or gods?

I presume you don't believe in almost all of the many gods listed in Chapter 1, or in the hundreds more that I didn't mention. Chapters 2 and 3 might have convinced you that holy books like the Bible and the Quran don't provide any good reason to believe in any gods. Chapters 4, 5 and 6 might have led you away from believing that religion is necessary for us to be good. But you might still cling to belief in some kind of higher power, some sort of creative intelligence who made the world and the universe and – perhaps above all – made living creatures, including us. I clung to such a belief myself until I was about 15, because I was so deeply impressed by the beauty and complexity of living things. Especially by the fact that living things *look* as though they must have been 'designed'. I finally gave up on the very idea of any gods when I learned about evolution and the true explanation for why living things look designed. That explanation – Charles Darwin's explanation – is as beautiful and subtle as the living things that it explains. But it takes time to develop. It will occupy most of Part Two of this book. But even that is not long enough to do justice to such a big subject. I hope it may interest you enough to lead you to other books on evolution.

PART TWO

Evolution and beyond

Surely there must be a designer?

Imagine a gazelle out on the African savanna, running for its life away from a sprinting cheetah, whimpering out what may well be its last breath. Perhaps, like me, you sympathize with the gazelle. But the cheetah has hungry cubs. If she can't catch prey she, and her cubs, will starve. Which might be a more unpleasant death than the gazelle's swift one.

If you've seen a film of a gazelle and a cheetah running – perhaps one of David Attenborough's documentaries – you've probably noticed how beautifully, how elegantly *designed* both animals seem to be. Both of these muscular, taut-sprung bodies have 'fast' written all over them. The top speed of a cheetah is around 100 kilometres per hour. That's about 60 miles per hour. Some reports even put the top speed as high as 70 mph, which is quite a feat when you have no wheels, only feet to propel you. And a cheetah can accelerate from 0 to 60 mph in three seconds, which is about what a Tesla (in 'insane mode') or a Ferrari can do.

The cheetah can't keep it up for long. Cheetahs are sprinters, unlike wolves, who are long-distance runners. Although their top speed is slower (more like 40 mph), wolves persevere and can eventually run their prey down.

Cheetahs need to stalk their prey until they are really close, close enough for a final, short sprint. Anything longer than a sprint exhausts them and they have to give up the chase. Gazelles can't run as fast as cheetahs (again, about 40 mph), but they 'jink' (dodge from side to side) which makes it hard for a sprinting cheetah to catch them – especially because, when you are sprinting at very high speed, it is hard to turn.

Like other antelopes, gazelles also 'pronk' when being chased. Pronking (or 'stotting') means leaping high into the air. This is surprising, because it must slow their progress and consume energy. It might be a signal to the cheetah: 'Don't bother to chase me, I'm a strong, fit gazelle who can leap high into the air. This probably also means I'm harder to catch than other gazelles. You'd be better off going for another member of my herd.' The gazelle doesn't think these arguments out. Its nervous system is just programmed to pronk, without understanding why. Whether by pronking or jinking, if a gazelle can evade capture for just long enough that the sprinting cheetah gets tired and has to stop, it's safe. For another day.

Both cheetahs and gazelles seem superbly 'designed'. The spine of the cheetah bends way, way back, and then thrusts the other way, almost bending double, powering the legs in a frenzied gallop. Its lungs are unusually large for an animal that size. So are the nostrils and air tubes, because of the need to get lots of oxygen into the blood fast. The heart, too, is especially large, to pump plenty of that oxygen-rich blood to the muscles, frantic with effort. But, quite apart from the *size* of the heart, the fact of having a heart at all, having this complicated four-chambered pump working away constantly, is remarkable enough. The mathematics of heart pumping has been cleverly worked out. I won't even try to explain it because it's too complicated for me to understand myself.

How did all this complexity come about? Must it have been designed by a mathematically minded genius? The answer is an emphatic, if surprising, no – and we'll see why in the following chapters.

Now think of the cheetah's eye, menacingly fixed on its prey while it alternately crouches and creeps stealthily forward. Or the gazelle's eye, restlessly scanning for lurking big cats. The vertebrate eye is a camera. A digital camera really because, instead of a film at the back, it has a *retina* with millions of tiny light-sensitive cells. We can call them photocells. Each photocell is connected, via a series of nerve cells, to the brain. There are several 'maps' of the retina in the brain. By 'map' I mean a corresponding pattern, so that cells next to each other in the brain are connected to photocells next to each other in the retina in the same orderly fashion, both side-to-side and up-and-down on the map.

The resemblance to a camera goes further. The pupil is widened or narrowed by special muscles attached to the iris (the coloured part of the eye). You can see this if you look at your own eyes in a mirror. Hold a torch pointing at your left eye, and then switch it on while looking at the right eye in the mirror. You'll see the pupil shrink. In an automatic camera, too, the 'iris diaphragm' (even the name comes from the eye) opens or closes just the right amount to let the right amount of light in. It shrinks the aperture when the sun comes out. Expands it when the sun goes in. Exactly like the iris in the eye. The pupil doesn't have to be round, like ours, by the way. Gazelle pupils are horizontal slits. Cat pupils are vertical slits in bright light, widening to circles when light levels are low. What matters is that the pupil, and the muscles surrounding it, control how much light gets into the eye. Incidentally, the image on the retina is upside down. Can you see why that doesn't matter? Why it doesn't mean the world looks upside down to us?

Again like a camera, an eye contains a lens that can be focused on near objects and then refocused on distant objects – or, of course, anywhere in between. Cameras and fish eyes do it by moving the lens back and forth. The eyes of cheetahs, gazelles, humans and other mammals do it in a less obvious way. They change the shape of the lens itself, using special muscles attached to the lens. Chameleons, which have independently swivelling eyes on little conical turrets, can focus the two eyes independently (using the fish/camera method, not the lens-squeezing method), and they judge the distance to a target, such as a fly, by measuring what they have to do to focus on it. The fly then doesn't know what hit it. In fact what hit it – at great speed – was the chameleon's tongue, which (amazingly) is longer than the chameleon itself, shooting out explosively like a sticky harpoon. The tongue harpoon is then reeled in, complete with the doomed insect stuck to the tip.

Chameleons and cheetahs have something in common. Both stalk their prey slowly and stealthily until they are close enough. Close enough for what? In the cheetah's case, for a final, explosive sprint. In the chameleon's case there is a kind of final sprint, too. But the sprint is by the tongue alone while the body stays rock steady. You remember the cheetah accelerates from 0 to 60 mph in three seconds? The chameleon's tongue has the equivalent of 300 times that acceleration. But it hits (or misses) the fly long before it actually reaches 60 mph. After all, the tongue is only (only!) slightly longer than the chameleon's whole body, so there isn't time to reach 60 mph, even at that phenomenal rate of acceleration.

Once again, this all looks as though it demands a designer, doesn't it? Once again, it really doesn't, as we'll see in the next chapters.

Exactly how the chameleon's tongue works has long been a bit of a mystery. One early suggestion was that it was inflated by hydraulic pressure, like an erecting penis only much faster. The hydraulic method is also used by jumping spiders (lovable little creatures which leap high into the air, having belayed themselves to the ground with a silk thread). Blood is pumped violently into the legs, which abruptly straighten and shoot the spider upwards. Butterfly and moth tongues work like that, too. They are coiled up at rest, then uncoiled by hydraulic pressure like a 'party horn' – one of those toys you blow into, and it shoots out into somebody's face, often making a blaring noise.

Although it's partly wrong, that hydraulic theory did get one thing right: the chameleon tongue is hollow. But instead of containing only fluid under pressure, it also contains a long, stiff, lubricated spike called the hyoid process. Obviously the tongue is much longer than the hyoid spike. So the resting tongue has to be accommodated in folds around the spike. Wrapped round and round are strong muscles. This fact naturally suggested the next theory of how the tongue works – again wrong, but closer to the truth. This was the theory that when the muscles contract around the hyoid spike the lubricated hollow tongue is squeezed outwards from its telescopic folds. Like when you squeeze an orange pip (seed) and it shoots off. That's almost what happens. But not quite.

The thing is, no muscle can contract fast enough to deliver the 'insane' acceleration of the chameleon tongue. For that sort of acceleration, the energy provided by the muscles needs to be *stored* ahead of time and then released later. That's how catapults work. And crossbows and longbows. Your arm muscles aren't capable of throwing an arrow very fast, but a bendy bow is. Your arm muscles slowly pull the bowstring back and the muscular energy is stored in the bending bow. Then the stored energy is suddenly released when your fingers let go, and the arrow shoots off much faster and more lethally than you could possibly throw it. The energy originally came from your muscles slowly pulling. The release of the energy is postponed and sudden: stored in the bow. In a catapult, the energy of your arm muscles is stored in the stretched elastic.

How does stored energy power the chameleon's tongue? The muscles around the hyoid spike do indeed provide the energy to shoot the tongue out. But, as with a catapult or bow, that energy is stored. It's stored in an elastic sheath which lies between the muscle and the well-lubricated hyoid spike. It's this elastic sheath, rather than the muscles themselves, which 'squeezes the orange pip' when finally the spring-loaded mechanism is suddenly released and the harpoon tongue shoots out: much faster because of the elastic sheath than it would be if the muscles squeezed the 'orange pip' directly.

The tongue is not sharp, like a harpoon. Instead, it has a sort of knob on the end. The knob is sticky and it has a suction cup. This sticks to the poor insect, which is then reeled into the chameleon's mouth by a different set of muscles called the retractor muscles. The knob is a relatively heavy projectile, whereas the rest of the tongue is more like a dangly rope. The knob travels 'ballistically' – which means that, once it has been launched, it's no longer under the chameleon's control. Just like the stone

1 How do religions start? Some are so recent, we can actually watch them emerge. On the island of Tanna in the South Pacific, Prince Philip has been revered as a deity since he visited nearly 50 years ago. Equally young are the cargo cults of several Pacific islands. If new religions can spring up so suddenly and rapidly in our own time, just imagine the scope for distorted legends to grow in the many centuries since the major religions of the world began. (See Chapter 3.)

2 (above): Speed written all over them. Did God design cheetahs to catch gazelles at the same time as he designed gazelles to escape? (See Chapter 7.)

3 (below): The chameleon's tongue is a beautiful natural harpoon. Note the hyoid bone inside the tubular tongue, which plays a central role in the harpoon's explosive speed. Elegant 'design'. Or is it? (See Chapter 7.)

Can you see the octopus (**4, top left**)? No, and nor could the photographer. It suddenly materialized, ghostly white (**5, top right**). How does a male squid (**6, left**) go white to scare away rivals while staying brown to reassure a female? Easy. Go two-tone. Was the flounder (**7, below**) designed by God? More likely designed by Picasso! In fact, the curious distortion of the head has evolutionary history to blame. No designer would ever have chosen this way to make a flatfish. (See Chapter 7.)

8 Naturally selected camouflage, every last detail honed to perfection by the sharp eyes of predators. You can see why people were tempted to credit the hand of God. (See Chapter 7.)

9 Look what selection can do. If (artificial) selection takes only 30 centuries to transform the wild plant *Brassica oleracea* (**top left**) into Brussels sprouts, cauliflower, cabbage and Romanesco (not to mention broccoli, curly kale, kohlrabi etc.), just think what (natural) selection could do in the 3 million centuries since our ancestors were fish. (See Chapter 8.

10 Two kinds of architecture. La Sagrada Família church (**near left**) was designed to the last detail, by a great architect. The termite castle (**far left**), photographed by Fiona Stewart in Australia, was not designed. Not by termites, not by their DNA, not by God. (See Chapter 10.)

Hard to believe the starlings (**11, above**) are not directed by a master choreographer, so perfectly coordinated are they. The flock looks like a single organism, a giant aerial amoeba. But there's no choreographer. A computer simulation (**12, below**) shows how it's done. (See Chapter 10.)

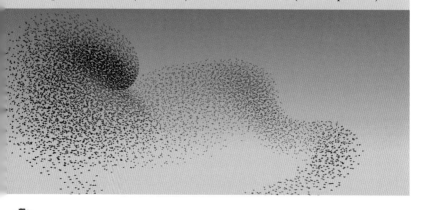

13 A trick picture. The top half is a real photograph of real galaxies. The bottom half is a computer simulation, the Illustris simulation of the development of the universe beginning almost immediately (a mere 300,000 years) after the Big Bang. Can you tell the difference? (See Chapter 12.)

from a catapult or the arrow from a bow. Or indeed a harpoon, which it more strongly resembles because, like the chameleon's tongue, it remains tethered to the launching apparatus. An intercontinental *ballistic* missile (ICBM) is so called because, once launched, it's on its own. As opposed to a *guided* missile, whose course is corrected while in flight, to help it home in on the target.

By the way, that same catapult trick, storing energy from slow muscles in quick-release elastic, is also used by jumping insects like grasshoppers and fleas. Their 'rubber' is a wonderful substance called *resilin*. Resilin is even more efficient than rubber as an elastic. That means that a higher proportion of the stored energy is available for eventual release. *Efficient* is a technical term meaning that little energy is lost as heat. Inevitably some is lost, according to the unbreakable laws of thermodynamics – but there's no space to deal with those laws here. Most spectacularly of all, the elastic storage 'crossbow' trick is used by mantis shrimps to pack a punch which is utterly astounding in an animal only a few centimetres long. A pair of front limbs have evolved to become hammers or clubs, which batter prey at a speed of 50 mph. The acceleration is equivalent to that of a bullet from a .22 pistol. And that – unlike the bullet – is under water! To repeat, it is achieved using elastically stored energy. Direct muscle power couldn't possibly achieve such speed.

There's a bit more to the story of the chameleon's tongue. For instance, the hyoid spike itself moves forward to help the flying tongue on its way. It's as if, bow in hand, you run towards your target like a fast bowler in cricket, and then launch the arrow while still running. But I've already probably said enough to make you think: 'Surely somebody must have designed the whole amazing apparatus?' Again, you'd be wrong. Why do I keep saying this, and saying that it will all be explained in later chapters? Because this chapter is setting up the problem of what needs to be explained. And it's a big problem. I don't want to make light of it, which is why I devote this whole chapter to the problem itself, before we even start on the solution. As we'll see, only evolution by natural selection is a big enough theory to solve such a big problem.

Although chameleons have wondrous tongues and swivelling turret eyes, they are even more famous for something else: their ability to change colour to match their background. A politician who keeps changing his mind to blend in with prevailing opinion is sometimes teased as a 'political chameleon'. In their colour-changing skill chameleons are equalled by some flatfish, like plaice. But both are massively outclassed by octopuses and their kin. Chameleons and flatfish change colour slowly, over a time-scale of minutes. Octopuses, squids and cuttlefish, collectively called cephalopods, change colour from second to second.

Cephalopods are about as close to aliens as anything you'll find on this planet. They have eight (octopuses) or ten (squids and cuttlefish) arms surrounding their beak of a mouth. The arms are capable of astonishing feats of finely controlled, continuously bendy movement, which is especially remarkable since they contain no skeleton. They are the only animals that have true jet propulsion, and they use it to swim backwards, especially in sudden escape. And – which is why they come into this chapter – they can change colour very fast and in highly complex patterns. Tantalizingly, the way they do it is similar to how modern colour televisions work.

Switch on your television and look closely at the screen with a powerful magnifying glass. Unless it's an old-fashioned type (which has horizontal lines), you'll notice that the whole screen is covered with millions of tiny coloured dots, called 'pixels'. Every pixel is either red, blue or green, and every pixel can be turned on or off, brightened or dimmed, under the control of the TV set's electronics. The pixels are too small to be seen when you're sitting back watching television. But every colour, however subtle, that you see from your sofa is made by some mixture of pixel brightnesses. If you examine with your lens a bright white part of the picture, you'll see that all three colours of pixels, red, blue and green, are brightly lit. In a red part of the picture – not surprisingly – only red pixels are brightly lit. Similarly for blue and green parts of the scene. Yellow is made by switching on the red and green pixels together, purple by mixing red and blue, brown by a more complicated mixture. Grey is like white, with all three colours switched on – but weakly. The electronic apparatus of the television makes the entire moving picture by rapidly controlling the brightness of every single one of the millions of pixels. Computer screens work in the same way.

And – wondrous to report – so does the skin of an octopus, squid or cuttlefish. Its whole skin is a living TV screen. The pixels are not controlled electronically, however. Instead, each pixel is a tiny bag of coloured pigment. There are three different colours, just like in the TV screen, except that they aren't red, blue and green, they're red, yellow and brown. But, as with TV pixels, the three types are independently controlled, to vary the patterns of colour over the skin surface.

The cephalopod pixels are much bigger than the TV screen ones. They're bags of pigment, after all, and you can't make bags that small. How are they controlled? Each bag lies inside an organ called a chromatophore. Fish have chromatophores too, but they work in a different way. In cephalopods, the wall of the bag is elastic (interesting how elasticity keeps cropping up). There are muscle cells attached to the chromatophore. The muscles are arranged like the arms of a starfish, except that there are about twenty arms instead of only five. When the muscles contract, they stretch the walls of the bag so that a larger area of pigment is splayed out, and the chromatophore takes on the colour of the pigment. When the muscles relax, the bag shrinks to a dot because of its elastic walls, so its colour becomes invisible from a distance. Because the colour-change is controlled by muscles, and the muscles by nerves, it's fast: about one-fifth of a second to change. Not as fast as a television screen, but a lot faster than a chameleon's skin, where the chromatophores are controlled by hormones – substances that travel, inevitably slowly, through the blood.

The muscle contractions tugging at the chromatophores are controlled by nerves, and the nerves are controlled by cells in the brain. Nerves are fast (although not as fast as the electronic components in a television). Theoretically, if we could hook up a squid's brain cells to a computer, we could play Charlie Chaplin movies on its skin. Nobody's ever done that, although the squid itself comes close, with lovely waves of colour-change like speeded-up clouds wafting across the sky. Dr Roger Hanlon of the Woods Hole Marine Biology Laboratory kindly read early drafts of this chapter for me. And when he read my Charlie Chaplin suggestion he told me this. He and some colleagues took a dead squid and hooked up a nerve in its fin to an iPod. Of course the fin couldn't hear, but the wire pulsed electricity in time with the music's strong beat, and this stimulated the chromatophore muscles. The result was pretty crazy, like a disco light show. Search for 'Insane in the Chromatophores' on YouTube.

The story of cephalopod colour gets better still. First, you need to know that there are two ways things can get colourful. One is by pigment (ink, dye, paint), which absorbs some of the colour out of the sunlight and reflects the rest. The other way is by what is called 'structural coloration' or 'iridescence'. Iridescence doesn't work by absorbing sunlight. It reflects it, and it produces colours that vary depending on the angle from which they're seen and the angle at which the light hits the surface. Soap bubbles with their wonderful shimmering rainbow colours (Iris was the Greek rainbow goddess) are iridescent, and you may have seen the same thing in thin layers of oil on water. Iridescence is how peacocks make their lovely colours. Also the shining blue tropical butterflies called morphos.

Well, squids don't miss a trick, and structural coloration is another of the tricks they don't miss. Underneath the chromatophores is another layer of so-called 'iridophores'. Iridophores don't change their shape like chromatophores, but they glisten colourfully like a morpho's wing. Often a shining blue or green, which the chromatophores, being red, yellow or brown, can't do. And some, though not all, of these iridophores can change their colour, too – and they do it in a different way from chromatophores.

The iridophores lie in a separate layer underneath the chromatophores. So they form a colourfully glowing background which may be covered up, to a greater or lesser extent, by the winking chromatophores above them. In addition to the chromatophores and iridophores, and in yet another layer below the iridophores, there are so-called leucophores. These are white. Like snowflakes, they are white because they reflect light of all wavelengths: not neatly and tidily like mirrors, but scattered in all directions.

What do cephalopods use their changing skin colours and patterns for? Mostly camouflage. They can manipulate their chromatophores almost instantaneously to mimic their background. This trick is visible in a lovely film, shot by Roger Hanlon while he was diving in the Caribbean Sea off Grand Cayman Island. Plates 4 and 5 show a pair of stills from the film. As Dr Hanlon swam towards a clump of brown seaweed, to his amazement and delight part of the 'seaweed' turned a ghostly, threatening white. This made it seem to 'emerge' from the background, at which point it emitted a cloud of dark brown ink to obscure the view of any would-be predator and swam off. It's well worth looking up the movie. Search for 'Roger Hanlon octopus camouflage change'.

What's especially remarkable is that cephalopods manage to mimic the colour of their background even though their eyes are colour-blind. How do they know what colour the background is? Nobody knows for sure, but there is suggestive evidence that they have some kind of seeing organs all over the skin, or at least in several patches of skin. These organs are not true eyes. They can't form images. It's more like having a retina distributed over the skin. And a retina is all they'd need to form a workable picture of the colour of the background.

Camouflage isn't the only thing for which cephalopods use their astonishing powers of colour-change. Sometimes they use them to threaten enemies, or to court a mate. In another piece of film footage, Roger Hanlon captured a species of squid that uses white to threaten rival males, and stripy brown to court females (see plate 6). In his film a male squid achieves the amazing feat of colouring his right side white, to ward off other males, while at the same time colouring his left side stripy brown to please the female by his side. It's well worth watching. Search for 'Roger Hanlon' 'Signaling with skin patterns' (note, it's the American spelling, 'signaling'). You can see the male change colour instantaneously. A few seconds later, the female moves to the other side of the male, and he reverses his colour accordingly so that she sees only his courtship pattern. Cephalopods can also change the texture of their skin, puckering it up in ridges, spikes or protrusions.

If you do a web search for 'animal camouflage' you will find hundreds more examples of creatures using spectacular (in one sense; the opposite in another) camouflage to protect themselves: spiders, frogs, fish, birds and, above all, insects (plate 8 shows a few examples). It's the attention to detail that is so shattering. Each one looks like the work of a sublimely skilled creative artist. And that word 'creative' brings me back to the main point of this chapter. Everything about an animal or plant, every detail of every one, looks overwhelmingly as though somebody designed and created it. And through the centuries people have – wrongly – given the credit to one or other of the countless gods we met in Chapter 1. Or to no god in particular but some unnamed creator.

For me, even more impressive than camouflage is the sheer complexity of living bodies. We got a taste of this with the eye. Your brain is even more amazing. It contains about 100 billion nerve cells – straggly branching tree-rooty things (see the illustration below) – wired up to each other in such a way that you can think, hear, see, love, hate, plan a barbecue, imagine a giant green hippopotamus or dream of the future.

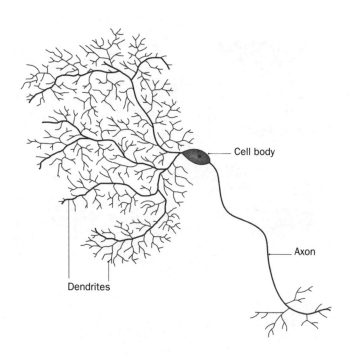

Cell body

Axon

Dendrites

On this page is a diagram of the chemical reactions that go on in a single cell of your body (you have more than 30 trillion cells altogether). The little blobs are chemical substances. The lines connecting them indicate chemical reactions between them. Don't bother with the detailed labels. But if the chemical reactions they indicate stopped, you'd die.

Now think of just one molecule from your body, haemoglobin. It's what makes your blood red, and it's vitally important for carrying oxygen from the lungs to wherever it's needed, for example the pounding leg

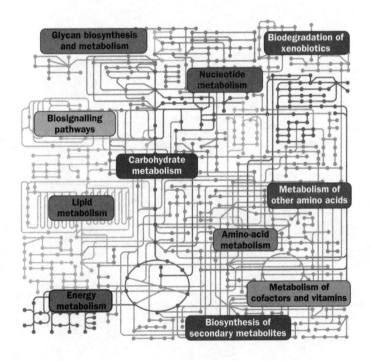

muscles of a sprinting cheetah or gazelle. More than six thousand million million million haemoglobin molecules are surging round in your blood at this moment. I once calculated for an earlier book (it seems a ridiculously high figure, but nobody has contradicted it) that haemoglobin molecules are springing into existence in a human body at a rate of four hundred million million every second, and others are being destroyed at the same rate.

Awe-inspiring complexity. Once again, it seems to demand a master designer. And once again, later chapters will show that it doesn't. That's quite a challenge; and the purpose of this chapter, to repeat it, is to show how big the challenge is. *Before* we step up to answer it.

Beauty raises the same kind of challenge. The glowing beauty of a peacock's tail – mostly achieved by structural, iridescent coloration – serves to attract peahens. We might even say it's beauty for beauty's sake. But beauty can also be 'functional': useful. I think airliners are beautiful, and their beauty comes from their streamlined shape. Flying birds are beautiful for the same reason. So are running cheetahs – although I don't suppose gazelles think so.

This chapter might have left you with the impression that living 'designs' are perfect. Not just beautiful but ideally fit for purpose, whether that purpose is seeing, changing colour, running fast to catch prey, running fast to avoid becoming prey, looking exactly like tree bark, looking irresistible to peahens or whatever. If it has, I have to disappoint you, just a little. Especially if you look under the skin of living things, you'll see flaws, and they are very revealing. What they reveal is evolutionary history. They are very much *not* what you'd expect to see if the animals had been intelligently designed. In fact, some are just the opposite.

Various species of fish make their living on the sea floor, and their bodies are flat. There are two ways of being flat. The obvious way is to lie on your belly and flatten the body from the top, so it spreads out sideways. That's what skates and rays have done. You could think of them as sharks that have fallen victim to a garden roller. But plaice, sole and flounders have done it differently. They lie on one side. Sometimes the left side, sometimes the right. But they never lie on the belly like skates.

It will have occurred to you that there's a problem with lying on your side if you're a fish. One of your eyes is against the bottom of the sea and is therefore pretty useless. That problem doesn't arise for skates and rays. Their eyes are on top of their flattened heads and both are useful for seeing things.

So, what did the plaice and flounders do about it? They grew a distorted, twisted skull, so that both eyes look upwards instead of one being flat against the sea bottom. And I do mean twisted and distorted (see plate 7). No sensible designer would have produced an arrangement like that. It makes no sense from a design point of view, but it has history written all over its Picasso-like face. Unlike the shark ancestors of skates and rays, the ancestors of these flatfish were shaped like a herring, a vertical blade. The left eye looked to the left and the right eye looked to the right. Symmetrically, as a good designer might wish. When they changed their way of life to live on the bottom, they couldn't go back to the drawing board, in the way that a designer would. Instead they had to modify what was already there. Hence the distorted head.

Here's another famous example of a revealing flaw: the retina of your eye. It's back to front. It's the same for all vertebrates. I've already described the retina as a screen of photocells. The photocells are hooked up to the brain by nerve cells. The sensible way to hook them up is the one used by cephalopods like octopuses. Their 'wires' connecting the photocells to the brain leave from the back of the retina in a sensible manner.

Not so the equivalent wires from the vertebrate retina. Here the photocells are wired backwards. Each photocell points away from the light. So how do the wires – the nerve cells – leading from the photocells manage to reach the brain? They travel over the surface of the retina, taking information from the photocells, and converge on a circular patch in the middle of the retina where they dive through and then head back to the brain (see the diagram overleaf). The place where they dive through is called the 'blind spot'. Because, not surprisingly, it is blind. What a ridiculous arrangement! The famous German scientist Hermann von Helmholtz (he was both a medical doctor and a pioneering physicist)

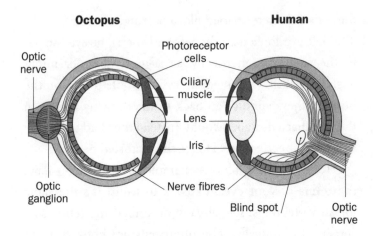

Octopus

Human

Optic nerve

Optic ganglion

Photoreceptor cells

Ciliary muscle

Lens

Iris

Nerve fibres

Blind spot

Optic nerve

once said that if a designer had presented him with the vertebrate eye he would have sent it back. Actually, although he would be perfectly justified in doing so, it works pretty well, as we can all see! The layer of nerve cells running over the surface of the retina is thin, and they are transparent enough to let light through.

My favourite example of bad design is the recurrent laryngeal nerve. The larynx is the voice box, in the throat. It's supplied by two nerves from the brain called the laryngeal nerves. One of these, the superior laryngeal, is sensibly wired up directly from the brain to the voice box. The other one, the recurrent laryngeal, is crazy. It goes down the neck from the brain, shoots straight past the larynx (the place where it is supposed to end up), way down into the chest. There it loops around one of the main arteries attached to the heart, then whizzes straight back up the neck and finally ends up in the larynx, where it should have stopped on the way down. In a giraffe, that's quite some detour. I saw this vividly when I assisted for a television programme in the dissection of a giraffe which had unfortunately died in a zoo.

Once again, this is obviously bad design, but it makes perfect sense if you look at the history. Our ancestors were fish. Fish have no neck. The fishy equivalent of the recurrent laryngeal nerve is not recurrent. It supplies one of the gills. The most direct route from the brain to the gill is behind the equivalent artery. It's not a detour at all. Later in history, when the neck started to lengthen, the nerve needed to make a slight detour. The neck got steadily longer as the generations passed. And the detour too got longer and longer. Even when the detour became absurdly long in the ancestors of the giraffe, because of the way evolutionary change works (as we'll see in the next chapter) it carried on just getting longer rather than changing the route altogether to jump over the artery. A designer would have taken one look at the nerve, as it passes within inches of the larynx on its way down the long, long neck, and said, 'Wait a minute, that's ridiculous.' Again, a Helmholtz would have sent it back. It's the same with the tube that carries our sperm from the testes to the penis. Instead of going by the most direct route, it travels up into the abdomen and loops over the tube carrying urine from the kidney to the bladder. Again, the detour makes sense only if you look at the evolutionary history.

I like the phrase 'History written all over us'. When we get cold, we get goosebumps. That's because our ancestors were hairy. When they got cold, each hair rose to thicken the layer of air trapped by the hairs that would keep us warm. Like putting on another sweater. We are no longer hairy all over our bodies. But the little hair-erecting muscles are still there. And they still – uselessly – respond to cold by raising non-existent hairs. Our hairy history is written all over our bare skin. Written in goosebumps.

To round off this chapter, I want to return to the cheetah and the gazelle. If God made the cheetah, he evidently put a lot of effort into designing a superb killer: fast, fierce, keen-eyed, with sharp claws and teeth, and with a brain dedicated to ruthlessly killing gazelles. But the same God put an equal amount of effort into making the gazelle. At the same time as he designed the cheetah to kill gazelles, he was busy designing the gazelle to be expert at escaping from cheetahs. He made both fast, so each could thwart the speed of the other. You can't help wondering, whose *side* is God on? He seems to be piling on the agony for both. Does he enjoy the spectator sport? Wouldn't it be horrible to think that God enjoys watching a terrified gazelle running for its life, then being knocked over and throttled by a cheetah gripping its throat so tightly that it can't breathe? Or that he likes watching a cheetah that fails to kill starve slowly to death, along with its pathetically whimpering cubs?

Of course, for an atheist none of that presents a problem because we don't believe in gods anyway. We are still at liberty to feel pity for the terrified gazelle or the starving cheetah and her cubs. But we don't find their situations difficult to explain. Darwinian evolution by natural selection explains it – and everything else about life – perfectly well. As we shall see in the next three chapters.

· 8 ·

Steps towards improbability

The previous chapter was filled with amazing examples of animals beautifully built, displaying uncannily perfect colour patterns, or doing apparently clever things to assist their survival. After each story, I asked: must there not have been a designer, a creator, a wise god who thought it all out and made it happen? What exactly is it about those examples – and you could tell similar stories for every animal and plant that ever lived – that makes people think there had to have been a designer? The answer is *improbability*, and I now need to explain what I mean by that.

When we say something is improbable we mean it's very unlikely to just happen by random chance. If you shake ten pennies and toss them on the table, you'd be surprised if all ten came up heads. It could happen but it's very unlikely. (If you enjoy arithmetic you might like to work out just how unlikely, but I'm content to say 'very'.) If somebody did the same thing with a hundred pennies it's still *just* possible they'd all come up heads. But it's so very very very improbable that you'd suspect a trick, and you'd be right. I'd bet everything I have that it was a trick.

With tossing pennies it's easy – well, straightforward, at least – to calculate the odds against a particular outcome. For something like the improbability of the human eye, or the cheetah's heart, we can't calculate it exactly just by using arithmetic, like we can with the pennies. But we can say that it's very very improbable. Things like eyes and hearts don't just happen by luck. It's this improbability that tempts people to think they must have been designed. And my task in this chapter and the next ones is to show that this thinking is mistaken. There was no designer. The improbability remains, whether we are talking about the improbability of an eye or the improbability of a creator capable of designing an eye. There has to be some other solution to the problem of improbable things. And that solution was provided by Charles Darwin.

For a living body, the equivalent of tossing pennies might perhaps be to scramble the bits of an eye, say, at random. The lens could end up at the back of the eye instead of the front. The retina could be in front of the cornea instead of behind the lens. The iris diaphragm could close when it's dark and open when it's light, instead of the sensible way round. Or open when you hear a trumpet and close when you smell an onion. The lens could be pitch black and not let any light through, instead of clear and transparent. Even having a retina or an iris diaphragm at all wouldn't happen if you scrambled the bits of *them* at random.

Or imagine a randomly scrambled cheetah. It could have all four legs on one side, so it keeps toppling over sideways. The rear legs could be stuck on backwards, so they gallop in the opposite direction to the front legs and the cheetah doesn't move either forwards or backwards but tries to tear itself in half. The heart could be connected to the windpipe, so it pumps air instead of blood. The cheetah could have teeth in its backside instead of in its mouth. And a totally scrambled cheetah wouldn't have legs or heart or teeth at all. It would be a jumbled mess: a puréed cheetah smoothie.

This is just silly, as I'm sure you realize. There's an infinite number of ways you could scramble the bits of a cheetah, and only a tiny number of them could run. Or see. Or smell. Or have babies. Or indeed stay alive. There's an infinite number of ways you could scramble the bits of a chameleon, and only a tiny number of them could shoot a tongue out at an insect. It's completely obvious that animals and plants do not come about by random chance. Whatever else is the explanation for cheetahs and gazelles, the lightning-fast chameleon tongue, the chromatophores and iridophores and leucophores of a squid, it cannot be random chance. Whatever is the true explanation for all the millions of animals and plants, it cannot be luck. We can all agree about that. So, what is the alternative?

Unfortunately, at this point many people go straight down the wrong path. They think the only alternative to random luck is a designer. If that is what you think, you're in good company. It's what almost everybody thought until Charles Darwin came along in the middle of the nineteenth century. But it's wrong, wrong, wrong. It isn't just a wrong alternative: it's no alternative at all.

The wrong argument was most famously expressed by the Reverend William Paley in his 1802 book *Natural Theology*. Imagine you're out for a walk on a heath, Archdeacon Paley said, and you happen to kick a stone. You are not impressed by the stone. It just happens to be there, and it just happens to have the rough, irregular, knobbly shape it has. A stone is just a stone. It doesn't stand out from all the other stones. But now, says Paley, suppose you stumbled over not a stone but a watch.

A watch is complicated. Open up the back and you see lots of cogwheels, springs, delicate little screws. (In Paley's time, of course, this wouldn't have been a modern digital wristwatch: it would have been a mechanical time-piece, a pocket watch with a beautifully and expertly crafted movement.) And all those tiny interlocking parts work together to do something useful: in this case, tell the time. Unlike the stone, the watch couldn't have just *happened* by luck. It had to have been deliberately designed and put together by a skilled watchmaker.

Of course, you can easily see where Paley was going with this. Just as the watch must have had a watchmaker, the eye must have had an eye-maker, the heart must have had a heart-maker. And so on. It's possible you are now even more persuaded by Paley's point than you were before. Even more reluctant to hear that it's wrong and that there really is no need for a creator god.

The scrambling argument shows that whatever else the explanation of the beautiful improbability of living things may be, it certainly can't be random luck. That's pretty much what improbability *means*. But now here's a little twist to the argument. It may be little, but it's very important: the Darwinian twist. Suppose that, instead of scrambling all the bits of a cheetah at random and making a horrible mess, we change just one little bit of the animal, again in a random direction. The key point is that we change it only a very small amount. Suppose a cheetah is born with claws just a tiny bit longer than in the previous generation. Now we don't have a horrible mess of scrambled cheetah. We still have a proper living, breathing, running cheetah. It has changed at random, but only very slightly. Now it is quite likely that this tiny change makes the cheetah a little bit worse at surviving. Or perhaps a bit better. Perhaps longer claws give the cheetah a better grip on the ground, and this helps it to run just that little bit faster. Like the spiked running shoes that athletes wear. So it catches a gazelle that would otherwise have narrowly escaped. Or perhaps the claws give the cheetah a better grip on the prey when it's caught, so it has less chance of wriggling free.

And how did that cheetah get its slightly longer claws? Somewhere in the cheetah genome, there is a gene that affects claw length. A baby cheetah always inherits its genes from its parents. But we're now talking about a new baby in which one gene, a gene which affects the claws, isn't quite the same as the parental version. It changed at random. The gene has 'mutated'. The process of mutation itself is random – it is not specifically guided towards improvement. Most mutant genes, in fact, make things worse. But some – as in our example of the slightly longer claws – happen to make things better. And in that case, the animals (or plants) that possess them are more likely to survive, and pass on their genes, including the mutant ones. That's what Darwin called natural selection (although he didn't use the word 'mutation').

A random mutation could make the claws blunter instead of sharper. And maybe less good at running or gripping prey. The smaller the change, the closer the probability gets to 50 per cent that it's an improvement. To see why, imagine that the change is very large. Say the mutant claws are a foot long. That's bound to make the cheetah less successful. It'll trip over its monstrous claws and they'll break when they try to grip anything. The same will be true of a big change *in either direction*. If the legs suddenly become two yards long or only six inches long, the cheetah will perish swiftly. Now think about a very small change, again in either direction. Imagine a mutation that is so small as to have almost no effect at all on the cheetah's body. A change like that will have hardly any effect on the animal's success, either way. A very small change, so small that it is almost – but not quite – zero, will have an approximately 50 per cent chance of being an improvement. The larger the mutation, in any direction, the greater the likelihood that it will damage the animal's performance. Large mutations are bad.

Small mutations approach a 50 per cent chance of being good.

Darwin realized that successful mutations are nearly always small. But the mutations that scientists study are usually large, for the obvious reason that small ones are hard to detect. And because large mutations, in any direction, are almost always bad, this has led some people to doubt evolution because they think all mutations are bad for survival. It may be true that all the mutations big enough to be easily studied in the lab are bad for survival. But it's the small ones that matter in evolution.

Darwin persuaded his readers of the power of selection by first pointing to domestication. Humans have changed wild horses into dozens of different breeds. Some, like carthorses and medieval chargers, are larger than wild horses. Others, like Shetland ponies and Falabellas, are much smaller. We (that is, our human ancestors) made carthorses, by choosing to breed from the largest individuals in successive generations. We made Falabellas by breeding from the smallest. Generation by generation, we made all the breeds of dogs from wolf ancestors. We made Great Danes and Irish Wolfhounds by breeding from the largest as the generations went by. We made Chihuahuas and Yorkies by breeding consistently from the smallest. Starting with the wild cabbage, which is an ordinary, nondescript wild flower, we made Brussels sprouts, cauliflowers, kale, broccoli, kohlrabi and the mathematically elegant Romanescos (see plate 9).

All were made by humans practising artificial selection. Farmers and gardeners, dog-breeders and pigeon-fanciers have known about the power of selection for centuries.

What Darwin brilliantly realized is that you don't need the human selector. Nature does the job all by itself, and has been doing it for hundreds of millions of years. Some mutant genes help animals to survive and reproduce. Those genes become more frequent in the population. Other mutant genes make it harder for them to survive and reproduce, and so become less frequent in the population until they disappear altogether. It only takes a few centuries to turn a wolf into a whippet or a Weimaraner. Just think how much change could be achieved in a million centuries. Since our ancestors were fish crawling out of the sea, three million centuries have gone by. That's an awful lot of time – a huge opportunity for change – step by step down the generations. The key point about mutation, to return to it, is that successful mutations, though random, are small. The mutant animal is not a randomly scrambled mess. Each random change makes it only a little bit different from the previous generation.

Let's go back to our cheetah to see how nature does the job of a farmer or gardener or dog-fancier. The cub with the mutant gene grows up, and its slightly longer claws help it to run just that little bit faster. So it catches more prey, which means that its cubs are better fed and are more likely to survive and have cubs of their own. Some of these new cubs – grandchildren of the mutant –

inherit the mutated gene, so they too grow up with slightly longer claws. They too run extra fast because of it and they too therefore have more cubs – great-grandchildren of the original mutant. And so on. It's as though a human breeder systematically chose the fastest individuals to breed from. But there is no human breeder. Survival does the job instead. You can see what's going to happen. As the generations go by, the mutated gene becomes more and more common in the population. Eventually a time comes when almost the entire population of cheetahs has the mutated gene. And they are all running just a little bit faster than their ancestors did.

This now puts extra pressure on the gazelles. Not all gazelles can run equally fast. None can run as fast as a cheetah, but some gazelles can run faster than other gazelles and they are the ones more likely to escape being eaten. This makes them more likely to survive to have babies. And their babies will inherit the genes for running fast. Genes for running slowly are more likely to end up in the bellies of cheetahs, lions or leopards, and consequently less likely to end up in future generations of gazelles. If, again by a random change in an existing gene, a new mutant gene were to arise which helps gazelles to run faster, it would spread through the gazelle population. Just like the cheetah mutation. It could be a change in the hooves. Or a change in the heart. Or some deeply buried change in the chemistry of the blood. The details don't matter here. If any gene helps gazelles to survive, by any means whatsoever, it'll get passed on to their children. So, like the cheetah gene, it will eventually spread until it becomes universal in the population. As the generations go by, both cheetahs and gazelles, hunters and hunted, have become just a little bit faster. We say there has been an *evolutionary* change on both sides.

I like the metaphor of the *arms race*. Of course, an individual cheetah and an individual gazelle literally run a race against each other. But that's not an arms race. That's just a race, and it ends rather swiftly in victory for either the cheetah (a meal) or the gazelle (escape). Arms races are run more slowly, in evolutionary time rather than individual cheetah/gazelle time. The arms race is between the gazelle species and the cheetah species (also lion species, leopard species, hyena species, Cape hunting dog species). And the result of the arms race is improvement, over the slow evolutionary time-scale. Improvement in *equipment* for survival: improvement in running speed as the generations go by; improvement in legs, stamina, dodging skill, sense organs to detect predators, or prey; improvement in blood chemistry to get oxygen to muscles fast.

Just like in human life, nothing is free. Improvements have to be paid for. Improved running speed demands longer legs with less heavy bones. And that is paid for in increased likelihood of broken legs. Human artificial selection has bred racehorses to run faster than natural selection ever did. But racehorses' long, slender legs are consequently more likely to break. Imagine what would have happened to wild horses, if they had been driven by the arms race against sabretooth tigers to run as fast as modern racehorses. The fastest individuals might have been more likely to outrun the sabretooth, with their longer legs and lighter bones. But they'd also have been more likely to break a leg. Then they'd have been easy meat for the sabretooth. So in practice we'd expect the arms race to lead to a compromise: wild horses would run fast, but not quite as fast as a human-bred racehorse. And that's what actually happened. Not surprisingly, modern racehorses often do break their legs. And tragically have to be shot.

And it's not just broken legs and the like that put limits on the arms race. Economic limits are also important. Fast running muscles are costly to make. You need food to turn into muscle. That food could have been put into something else: into making milk for babies, for instance. Human arms races are also economically costly. The more money you put into bombers, the less money is available for fighters. Not to mention less money for hospitals and schools.

Think of the economic calculation that a plant, such as a potato, has to do. A plant is a good example, because while we might be tempted (wrongly) to think that a gazelle or a cheetah or a horse does calculations in its head, nobody could seriously imagine that a plant does sums. And doing calculations consciously is exactly what we are *not* talking about. The equivalent of calculations is done by natural selection over the generations. So, back to the potato plant. It has a limited amount of 'money' to play with. 'Money' here means the energy resources that ultimately come from the sun, turned into the currency of sugar and often stored as starch, for instance in a potato tuber. The plant needs to spend some money on leaves (to take in sunlight to make yet more money). It needs to spend some money on roots (to take in water and minerals). It needs to spend some money on underground tubers (to store money for next year). It needs to spend some money on flowers (to attract insects to pollinate other potato plants and spread the genes – including genes for getting the spending decisions right). Potato plants that get their 'calculations' wrong – perhaps not spending enough on tuber storage for next year – are less successful in passing on their genes. As the generations go by, plants that get their economic sums wrong become less numerous in the population. And that means that genes for getting economic sums wrong become less numerous. The population 'gene pool' becomes more and more full of genes for getting the economic sums right.

Having learned from the potato plant that we are not talking about conscious calculations, we can safely go back to gazelles and talk about how they get their economic balance right. The details are different from the potato but the principles are the same. Gazelles need to be cautious of cheetahs and lions. They need to be scared.

They need to keep a watchful eye open. And a 'watchful' nose, for they often use smell to detect danger. But, importantly, they also need to spend a lot of time eating. Weight for weight, plant food is less nutritious than meat, so a herbivore – an animal that eats only plants – like a gazelle or a cow needs to keep eating almost all the time. A gazelle that was too scared would keep running away on the slightest suspicion of danger and wouldn't have enough time to eat. On the African plains you can sometimes see antelopes and zebras grazing within sight of lions, knowing full well they are there. They keep a wary eye open in case the lions show signs of starting a hunt. But they go on grazing. Over the generations, natural selection has achieved a fine balance between being too scared (and therefore not getting enough to eat) and not being scared enough (and therefore getting eaten).

Evolution consists of changes in the proportions of genes in populations. What we *see* from outside is changes in bodies or behaviour as the generations go by. But what is really going on is that some genes are becoming more numerous in the population and others less numerous. Genes survive, or fail to survive, in the population as a direct result of their effects on bodies and behaviour, only some of which are visible to us. It's not just cheetahs and gazelles, zebras and lions; it's chameleons and squids, kangaroos and kakapos, buffaloes and butterflies, beech trees and bacteria, every animal and plant, every mushroom and every microbe – they all contain the genes that helped an unbroken line of ancestors to survive and pass those genes on.

You and I and the Prime Minister, your cat and the birds singing outside your window, every single one of us can look back at our ancestors and make the following proud claim: not a single one of my ancestors died young. Plenty of individuals died young, but they are not the ones that became ancestors. Not a single one of your ancestors fell over a cliff, or was eaten by a lion, or died of cancer, before living long enough to have at least one child. Of course that's obvious when we think about it. But it's really, really important. It means that every single one of us, every animal and plant and fungus and bacterium, every one of the seven billion people around the world, contains genes for being good at surviving and becoming an ancestor.

The details of what makes us good at surviving vary from species to species. For cheetahs it's sprinting, for wolves it's long-distance running, for grass it's being good at absorbing sunlight and not minding too much about being cropped by cows (or lawnmowers), for cows it's being good at digesting grass, for hawks it's being good at hovering and spotting prey, for moles and aardvarks it's being good at digging. For all living creatures, it's getting the economic balance right. It's being good at thousands and thousands of things, all working together through every corner and cranny of the body, through every one of billions of cells. The details vary hugely, but they all have one thing in common. They are all ways of being good at passing on genes to future generations. Passing on genes that make them good at surviving and passing the same genes on. Just different detailed ways of doing the same thing: surviving and passing on genes.

We agreed that an eye or any organ that's complicated (like Paley's watch) is too *improbable* to have just happened (like Paley's stone). An excellent seeing device like a human eye cannot spring spontaneously into existence. That would be too improbable, like throwing a hundred pennies down and getting all heads. But an excellent eye can come from a random change to a slightly less excellent eye. And that slightly less good eye can come from an even less good eye. And so on back to a really rather poor eye. Even a very, very poor eye is better than no eye at all. You can tell the difference between night and day, and perhaps detect the looming shadow of a predator. And the same kind of thing is true not just of eyes but of legs and hearts and tongues and feathers and blood and hair and leaves. Everything about living creatures, no matter how complicated, no matter how improbable – as improbable as Paley's watch – can now be understood. Whatever it is that you're looking at, it didn't spring into existence all at once. Instead, it came from something just a little bit different from what went before. Improbability dissolves away when you see it as arriving *gradually*, stealthily, step by tiny step, where each step brings about only a really small change.

And the first step may not have brought about anything very good at all.

Improbable things don't jump into the world suddenly. As I said before, that's what improbable means. Paley was right about the watch. A watch can't spring spontaneously into existence. It has to have a watchmaker. Watchmakers, too, don't spring spontaneously into existence. They are born as complicated babies: human babies that grow into human adults, with human hands and brains and ability to learn a skill like watchmaking. Those human hands and brains evolved gradually from ape hands and brains; those apes evolved gradually from monkey-like ancestors; they in turn evolved by gradual, slow, painfully slow degrees from shrew-like ancestors; from fish-like ancestors before that; and so on. It was all gradual, slow, never sudden, never improbable like a watch spontaneously springing into existence at one go.

Designers need an explanation, just as watches do. Watchmakers have their explanation: being born from a woman, and before that by slow, gradual evolution through a very long chain of ancestors – the same explanation as for all living things. So where does that leave God, the alleged designer of everything? If you don't think about it very hard, God seems to be a good explanation for the existence of improbable things like chameleons and cheetahs and watchmakers. But if we think about it more carefully, we can see that God himself is even more improbable than William Paley's watch. Anything clever enough – complicated enough – to design things has to arrive late in the universe. Anything as complicated as a watchmaker must be the end product of a long, slow climb from earlier simplicity. Paley thought his watchmaker argument established the existence of God. But, when properly understood, the very same argument goes in exactly the opposite direction: in the direction of disproving God's existence. Little did Paley know that he was eloquently and persuasively shooting himself in the foot.

9

Crystals and jigsaw puzzles

Let's go back to Archdeacon Paley's watch and look more carefully at how it differs from his stone. You can do the scramble test on both. If you take a particular stone and scramble the bits a thousand times, you'd need a lot of luck to hit upon exactly the same stone again. So you might say the stone is as improbable as the watch. But all those randomly cobbled stones will still just be stones and there'll be nothing special about any of them. Not so the watch. If you scramble the parts of the watch a thousand times, you'll get a thousand random messes. But not one of them will tell the time or do anything useful (not unless your random scrambling is ridiculously lucky!). They won't even be beautiful. That's the key difference between watch and stone. Both are equally improbable in that they are a unique combination of parts which won't just 'happen' by sheer luck. But the watch is unique in another, and more interesting way which separates it from all the random scramblings: it does something useful; it tells the time. Stones don't have that kind of uniqueness. There is nothing to single out any one of those thousands of randomly scrambled stones from all the rest. They're all just stones. Of all the thousands of ways the bits of a watch could come together, only one of those ways will be a watch. Only one will tell the time.

But now suppose, on your walk across the heath with Archdeacon Paley, you stubbed your toe on this:

Would you now be happy to say that this 'just happened', like Paley's stone? I suspect not. I think you – and certainly Paley – might be tempted to think it was carefully made by a designer, an artist. It wouldn't look out of place in a posh gallery, would it? A valuable work of art, fashioned by a famous sculptor. The shiny cubes seem so perfect, tastefully mounted in the rough stone base. For me it was a bombshell to discover that nobody crafted these beautiful objects. They just happened. Exactly like Paley's stone. Indeed, they *are* a kind of stone.

They're crystals. Crystals just grow, spontaneously. And some grow into precise geometric shapes which look, overwhelmingly, as though an artist had made them. These happen to be crystals of iron disulphide. There are many other crystals, spontaneously formed from different chemicals, which also look beautiful. Some are so beautiful – diamonds, rubies, sapphires, emeralds – that they command fabulous prices, and people wear them around their necks or on their fingers.

To repeat: nobody carved that beautiful iron disulphide 'sculpture'. It just happened. Just grew. That's what crystals do. Crystals of iron disulphide are called pyrite, or sometimes 'fool's gold' because of their shiny colour. People who have dug them up have been fooled into thinking they were real gold and danced with joy, only to have their hopes cruelly dashed.

Crystals have pretty, geometrically precise shapes because their shape comes straight from the arrangement of their atoms. When water is cold enough it crystallizes into ice. The molecules in ice take up orderly positions next to each other. Like soldiers on parade, except that there are billions and billions of soldiers in even a small crystal: rank upon rank stretching off into the far distance in all directions. Unlike with soldiers, 'all directions' includes the up/down direction. The three-dimensional parade of molecules is called a lattice. Diamonds and other precious stones are also crystals, each with its own lattice pattern. Rocks, stones and sand are made of crystals, too, but often the crystals are so small and packed together that you can't easily see them as separate crystals.

Crystals also form in another way: when a substance is dissolved, usually in water, and the water evaporates. You can easily do this with ordinary table salt, sodium chloride. Boil a cupful of salt in water to dissolve it, then leave the solution to evaporate in a wide, shallow dish. As the days go by, you can see new salt crystals forming in the water. Crystals of common salt can be cubes like iron pyrite, or larger structures built of cubes and looking like four-sided ('ziggurat') pyramids. What happens is that sodium and chlorine atoms recognize each other and link arms. The proper name for the 'arms' is *bonds*. (Actually, in this case they're technically not atoms: they're *ions*, sodium and chloride ions, but the difference isn't important here.) Now, here's how crystals grow. Sodium and chloride ions still floating around in the water happen to bump into an existing crystal. They recognize the chloride or sodium ions already there on the edge of the crystal and link arms with them – and that's how the crystal grows. The reason crystals of common salt have square sides is that the 'arms' of the ions are at right-angles to each other. The crystal gets its shape from the right-angles of the files of 'soldiers on parade'. Not all crystals are square-sided, and you've probably already guessed why. Their 'arms' point at angles other than right-angles, so their 'soldiers on parade' line themselves up at those other angles. That's why fluorite crystals, for example, are octahedral – eight-sided.

Crystals can be large single stones with a nice geometric shape like a cube or an octahedron. But sometimes small crystals stick to each other to form more complicated shapes. The interior of each of the small building blocks of these complicated shapes betrays the underlying 'parade-ground of soldiers'. But the 'buildings' are more elaborate. Snowflakes are an example. You've probably read that no two snow-flakes are the same. In water ice the number of 'arms' is six, so the natural shape of each tiny ice crystal is six-sided. A snowflake is not just one of those tiny crystals. It's a 'building', made of lots of tiny six-sided 'bricks'. You'll notice that the six-sided design is reflected in the shape of the 'building', as well as the shape of the bricks themselves. Every snowflake has six-way symmetry (the illustration opposite shows a few exam-ples). But they are all different, and many of them are very beautiful.

It's worth pondering why snowflakes are all unique. It's because each has a unique history. Unlike crystals of salt, which grow at their margins in liquid water, snowflakes grow at their margins by adding tiny water crystals to the 'building' as they fall through clouds of water vapour. There are two ways in which they can grow. Which of the two predominates depends upon the 'micro-climate' in each tiny bit of cloud – how cold it is and how humid. Different micro-climates in the cloud vary in both temperature and humidity. Every snowflake experiences lots of different micro-climates as it floats down through the cloud: a unique moment-to-moment pattern of humidity change and temperature change. So the assembly of the 'building' follows a unique pattern and that particular snowflake ends up with a unique shape. It's a kind of fingerprint of moment-to-moment history.*

And what makes them beautiful? As with the image in a kaleidoscope, it's symmetry. All six sides, all six

---

* I owe my understanding of snowflakes to Brian Cox's beautiful book *Forces of Nature* (London, Collins, 2018).

corners, all six points or sets of points, are symmetrical. And why are they symmetrical? Because they are so small that all parts of the growing 'building' experience the same 'historical' pattern of humidity and temperature changes. By the way, although all snowflakes are unique, some are less beautiful than others. It's the beautiful ones that get pictured in books.

If we didn't know better, we might have thought, 'Oh look, snowflakes are so beautiful, and all unique. They must have been designed by a gifted creator with an ever-fertile mind able to think up so many millions of different designs.' But, as we have just seen, snowflakes and other beautiful crystals are like Paley's stone, not like Paley's watch. Science gives us a full and complete explanation of their beautiful and complex symmetry, and it also explains why they are all unique. Like Paley's stone, snowflakes 'just happened'. When molecules – or things generally – spontaneously form themselves into particular shapes like this – when they 'just happen' – the process is called self-assembly. I think you can see why. Self-assembly is very important in living things, as we'll soon see. This chapter is about self-assembly in life.

My champion example of living self-assembly is pictured on the title page of this chapter. It's a virus, the lambda bacteriophage. All viruses are parasites and this one, as the name 'bacteriophage' suggests, attacks bacteria. I think you'll agree that it looks like a lunar lander. And it behaves like one, landing on the surface of a bacterium where it stands, firmly mounted on its 'legs'. It then punches a hole through the bacterium's cell wall and injects its genetic material, its DNA, via its central 'tail' – which might better be called its 'hypodermic'. The machinery inside the bacterium can't tell the difference between the virus DNA and its own. It has no choice but to obey the instructions coded in the virus's DNA, and they tell it to manufacture lots more viruses which then burst out to land on, and re-infect, more bacteria. But what's interesting for this chapter is that the virus's 'body' self-assembles like a crystal. Or like a set of crystals. The head really looks like the sort of crystal you could wear round your neck (except it's much too small). It, and all the other parts of the virus, self-assemble just like crystals, from molecules drifting about inside the bacterium and slotting into the already growing crystal.

When I started talking about crystals, I used the metaphors of 'soldiers on parade' and 'linking arms'. We're now going to need a slightly different metaphor: a jigsaw puzzle. You could think of a growing crystal as an unfinished jigsaw puzzle. Just as a jigsaw might, it grows outwards from the middle, as pieces are added to the edges. But unlike the ordinary flat puzzle that sits on a table, a crystal is a three-dimensional jigsaw puzzle.

Around the unfinished puzzle, floating in the liquid, are thousands of jigsaw pieces. These might be sodium and chloride ions floating in water. Every time one of the floaters bumps into a crystal, it finds the correctly shaped 'hole' and slots itself in. So that's another way of picturing how a crystal grows at the margins. Now we're going to use the jigsaw metaphor to talk about what goes on in living creatures. In particular we're going to look at *enzymes*. We'll see what enzymes are in a moment.

Remember the picture in Chapter 7 of the chemical reactions going on in a cell: that enormously complicated spaghetti-junction of arrows and blobs? You might wonder how all those different chemical reactions can go on in the same tiny space, inside the same cell, without interfering with each other and getting muddled up. Suppose you went into a chemistry lab, grabbed all the bottles off the shelves and tipped them, all at once, into a great vat. You'd get a horrible mess – and maybe set off a lot of horrible reactions, even explosions. Yet somehow, in the cells of a living creature, lots of chemicals manage to stay separate without interfering with each other. Why don't they all react with one another? It's as though each one was in a separate bottle. But they aren't. How does it work?

Part of the answer is that the interior of the cell is not a single vat. It is filled with a complicated system of membranes, and these can act rather like the glass walls of test tubes. But that's not the whole story. There's something more interesting going on. And this is where enzymes come in. Enzymes are *catalysts*. A catalyst is a substance which speeds up a chemical reaction without actually being changed itself. It's a kind of fast-working miniature lab assistant. Catalysts can sometimes make a chemical reaction go millions of times faster, and enzymes are especially good at this. All those chemicals, muddled up together, don't react with one another unless there's a catalyst present: and it has to be a particular catalyst for each reaction. Particular reactions are turned on only when they are needed, by adding the right enzyme. You might think of an enzyme as a switch, which can be on or off, almost like an electric switch. Only when a particular enzyme is present in a cell is its one particular chemical reaction switched on. Even better, enzymes can 'switch on' other enzymes. You can see how elegant control systems could be built up with switches switching on (or off) other switches.

We know, at least in outline, how enzymes work. This is where the jigsaw puzzle idea comes in. Think of all the hundreds of molecules whizzing about in the cell as jigsaw puzzle pieces. Molecule X needs to find molecule Y in order to join up with it and make XY. The X/Y marriage is just one of the hundreds of vitally important chemical reactions in Chapter 7's 'spaghetti' diagram. There's a chance that an X will happen to bump into a Y. There's a smaller chance that they'll happen to bump at just the right angle to slot in and combine together. That happens so seldom that the rate at which XY is formed is extremely slow – so slow that if left to chance it would almost never happen. (This reminds me of my very first school report, when I was seven: 'Dawkins has only three speeds, slow, very slow and stop.') But there's an enzyme whose particular job is to speed up the rate at which Xs combine with Ys. And in the case of many enzymes, 'speed up' is an understatement. Again, the process works using the jigsaw principle.

An enzyme molecule is a great big complicated lump, with bulges and crevices all over its surface. When I say 'great big', it's only big by molecular standards. By the standards we're used to in our everyday lives it is tiny, too small to be seen by a light microscope. Let's take the particular case of the enzyme that speeds up our 'XY' chemical reaction. Among the crevices in its surface is an X-shaped hole which just happens to be right next to a Y-shaped hole. This is why it is a good 'lab assistant', specifically good at speeding up the X/Y combination. An X molecule falls, jigsaw-style, into the X-shaped hole. A Y-shaped molecule falls, jigsaw-style, into the Y-shaped hole. And, since the two holes are next door to each other in exactly the right way, the X and the Y find themselves snuggled together at exactly the right angle to combine with each other. The newly formed XY combination then pops out and floats away into the cell, leaving the two precisely shaped holes free to do the same thing with another X and another Y. So the enzyme molecule can be seen not just as a lab assistant but as a kind of factory machine, churning out XY molecules, using a steady stream of Xs and Ys as raw material. And, in that cell and in other cells elsewhere in the body, there are other enzymes, each one perfectly shaped – that is, with the right 'crevices' or 'dents' in the surface – to speed up other chemical reactions. I must stress that my language of 'crevice' and 'shape' constitutes a great oversimplification, but I'll stick with it because it's helpful for the purposes of this chapter. 'Shape' can mean not just physical shape but chemical affinity.

There are hundreds of enzymes, each one shaped differently, each one shaped to speed up a different chemical reaction. But in most cells only one or a few of the available enzymes are present. Enzymes are the main (though not the only) answer to the riddle of why the chemical reactions don't all happen at once, and don't all interfere with each other.

Enzyme molecules, then, sound like magic. Just as a cheetah's legs are beautifully shaped to run fast, enzymes are beautifully shaped to speed up particular chemical reactions. Just one special chemical reaction per enzyme. How do they get their beautiful shape? Are they carved into shape by a divine molecular sculptor? No. They come into being by a more complicated version of what growing crystals do. It's self-assembly again.

Every protein molecule is a chain of smaller molecules called amino acids. There are lots of different kinds of amino acids, but only 20 of them are found in living things. They all have names, and I could write out the 20 names, but let's not bother with the details. There are 20, and that's all we need to know here. Each protein molecule is like a necklace with amino acids for beads (a necklace with the clasp unfastened, not a closed loop). Proteins differ from each other in the exact sequence of beads from which each is made, all taken from the repertoire of 20 kinds of amino acids: 20 kinds of beads.

You remember that salt crystals grow when jigsaw pieces floating in water recognize their 'opposite numbers' at the edge of the crystal and slot themselves in. Well, think of the beads in the protein necklace as a selection from 20 kinds of jigsaw pieces. And some of them slot into other jigsaw pieces *somewhere along the same chain*. The result of this self-jigsawing, happening in various places all along the chain, is that the chain folds into its special shape. Like a piece of string tying itself into a very particular knot.

Now, I described an enzyme molecule as a complicated lump with bulges and crevices. That doesn't sound like a chain, does it? But it is. The thing is, any chain of amino acids has a tendency to fold itself into a particular three-dimensional shape. As I said, it's a bit like tying itself into a knot. The 'lump with knobs and crevices' is the knotty shape into which the chain assembles itself. Links in the chain are attracted to other particular links in the chain and stick to them, jigsaw-wise. And these hook-ups help to ensure that every instance of a particular chain folds itself into the same shape with the same bulges and crevices.

Actually that's not always quite true – and the exceptions are interesting. Some chains can tie themselves in one of two alternative knots. That can be extremely important, but I'm going to leave it aside here because this chapter is already complicated enough. For our purposes, we can think of each protein molecule as a chain of jigsaw pieces (amino acids) which folds itself into a very particular shape. The shape really matters, and it is determined by the particular sequence of amino acids and their tendency to slot, jigsaw-wise, into other amino acids in the same chain.

Here I can't resist a little story, which may seem unconnected but throws an interesting light on this idea of jigsaw pieces slotting in. It's about our sense of smell. Imagine the smell of a rose. Or of honey. Or onions. Apples. Strawberries. Fish. A cigar. A stagnant marsh. Every smell is different, unmistakable: beautiful or horrible, smoky or fruity, fragrant or foul. How is it that molecules, borne on the air into our nose, give rise to this smell or that smell? The answer is jigsaws again. The lining of your nose has thousands of differently shaped molecular crevices, each one just waiting for a molecule of exactly the right shape to slot in. A molecule of, say, acetone (nail-varnish remover) fits snugly into an acetone-shaped crevice, just like in a jigsaw puzzle. The acetone-shaped crevice sends a message to the brain saying, 'My kind of molecule has just slotted in.' The brain 'knows' that this particular crevice is an acetone-shaped crevice, so the brain 'thinks':

Aha, nail-varnish remover. The smell of a rose, or of a fine vintage wine, is made up of a complex mixture of jigsaw molecules, not just one as with acetone. But the point is the same: it's the molecular jigsaw principle at work.

Back to the main story. We've seen that the sequence of amino acids in the 'necklace' is responsible – through 'self-assembly jigsawing' – for the lumpy crevicy shape of the protein 'knot'. And we've seen that the crevices in turn are responsible for the protein's particular role as an enzyme, speeding up – usually so much that it amounts to switching on – a particular chemical reaction. There are lots of chemical reactions that could be going on in a cell at any one time. The ingredients are all there, ready to go. All each one requires is the right enzyme. And there are lots of enzymes that could be there, but only one is. Or only a few. So *which* enzymes are present is utterly crucial. They determine what the cell does. What the cell *is*, indeed.

So now you must be asking yourself, what determines the sequence of amino acids in the necklace of any particular enzyme, and therefore the lumpy shape into which the chain folds itself? That's obviously a hugely important question because so much else depends on it.

And the answer is: the genetic molecule, DNA. An answer whose importance is impossible to exaggerate. Which is why I've given it a paragraph to itself.

Like a protein molecule, DNA is a chain, a necklace of jigsaw pieces. But here the beads are not amino acids, they are chemical units called nucleotide bases. And there aren't 20 different kinds, only four. Their names are shortened to A, T, C and G. T jigsaws only with A (and A only with T). C jigsaws only with G (and G only with C). A DNA molecule is an immensely long chain, much longer than a protein molecule. Unlike a protein chain, the DNA chain doesn't tie itself into a 'knot'. Instead, it stays as a long chain – actually two chains jigsawed together in an elegant spiral staircase. Each 'step' of the staircase is a jigsawed pair of bases, and there are only four kinds of step:

A–T
T–A
C–G
G–C

The sequence of bases carries information, in the same way (almost exactly the same way) as a computer disc. And the information is used in two completely different ways: the genetic way, and the embryological way.

The genetic way is just copying. By a rather complicated version of jigsawing, the entire staircase is copied. This happens when cells divide. The embryological way is amazing. The code letters are read in triplets – three at a time. There are 64 possible triplets of four ($4 \times 4 \times 4 = 64$), and each of those 64 triplets is 'interpreted' either as a punctuation mark or as one of the 20 amino acids that go into making protein chains. When I say 'read', there is, of course, nobody to do the reading. Once again, it's all done automatically using the jigsaw principle. I'd love to go into the details, but that isn't what this book is about. For our purposes, what matters is that the sequence of the four types of bases in a stretch of DNA, when read in threes, determines the sequence of the 20 types of amino acids in a protein chain. The sequence of amino acids in a protein chain then determines how that protein chain coils up into a 'knot'. The shape of the 'knot' (its 'crevices' and other things) determines how it works as an enzyme, and therefore which particular chemical reaction it switches on in a cell. And the chemical reactions in a cell determine what sort of cell it is and how it behaves. Finally – and this is perhaps most wonderful of all – the behaviour of cells working together in an embryo determines how the embryo develops and turns into a baby. So it was ultimately our DNA that determined how each one of us developed from a single cell into a baby, and then grew into what we are now. This is the subject of the next chapter.

10

Bottom up or top down?

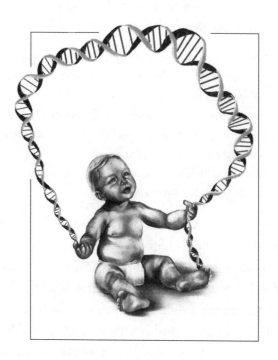

A great scientist – and larger-than-life character – of the twentieth century, J. B. S. Haldane, was once giving a public lecture. Afterwards, a lady stood up and said something like this:

> 'Professor Haldane, even given the billions of years that you say were available for evolution, I simply cannot believe it is possible to go from a single cell to a complicated human body, with its trillions of cells organized into bones and muscles and nerves, a heart that pumps without ceasing for decades, miles and miles of blood vessels and kidney tubules, and a brain capable of thinking and talking and feeling.'

Haldane gave a wonderful reply: 'But madam, you did it yourself. And it only took you nine months.'

The lady could have retorted, 'Ah, but my nine months as a developing baby were orchestrated by the DNA my parents gave me. I didn't have to start from scratch.' That is, of course, true. And her parents got the DNA from their parents, who in turn got it from their parents and so on back through the generations. What was happening during all the billions of years of evolution was that the DNA instructions for how to make babies were being gradually built up. Built up – honed and improved – by natural selection. Those genes that were good at making babies got passed on, at the expense of the genes that weren't. And the kind of babies that were made was changing, ever so gradually and slowly, over the millions of generations.

There's a rather charming hymn, 'All things bright and beautiful'. Perhaps you know it. It praises God for the detailed beauty of his creations, especially living creatures:

> He made their glowing colours
> He made their tiny wings.

But even if you believe God had something to do with creating animals, you'll realize that he didn't *directly* make glowing colours. Or wings, tiny or not. Wings and glowing colours, and all the other bits of a living body, develop anew, from a single cell by means of the processes of embryonic development. And embryonic development is supervised by DNA, via enzymes, which are made in the way we saw in the previous chapter. If God made glowing colours or fashioned tiny wings, he did it by manipulating the development of an embryo. Nowadays we know that means manipulating DNA (which then manipulates protein, and so on, in ways outlined in Chapter 9). And if – which is actually true – it is natural selection that (indirectly) paints those glowing colours, and fashions those tiny wings, natural selection too does it via DNA. DNA supervises the development of bodies, and DNA in turn is 'supervised' over many generations by natural selection. So, indirectly, natural selection 'supervises' the development of bodies.

You may have heard that DNA is a 'blueprint' for a body, but that's deeply wrong. Houses and cars have blueprints. Babies don't. The difference is entirely separate from the fact that cars and houses are designed whereas babies aren't. Here's the deeper difference. In a blueprint there's a one-to-one 'mapping' between each bit of the house (or car) and each bit of the blueprint. Neighbouring bits of house correspond to neighbouring bits of blueprint. If the blueprints of a house have been lost, you can redraw them simply by taking meticulous measurements of the house and drawing out a scaled-down version on paper. I've just had that done for my house. A man came with a laser gun to measure every room, and it only took him a couple of hours to draw out a complete plan, good enough to build an exact replica of my house.

You can't do that with a baby. There's no one-to-one mapping between points on a DNA 'blueprint' and points on a baby. In theory there could have been – it's not a totally silly idea. The plans of my house, carefully reconstructed by measuring every room, could be digitized in a computer. A modern genetics laboratory is capable of turning any computer information into DNA code, and that could include the digitized plans of my house. You could put the DNA in a test-tube and send it to another genetics lab, in Japan for example, where they could read the DNA and print out a faithful copy of the draw-ings. An exact replica of my house could then be built in Japan. Maybe on some other planet something like that happens when parents transmit their genetic informa-tion to their children: the parental body is 'scanned' and turned into a blueprint, which is then digitized in DNA (or that planet's equivalent of DNA). The digitized scan is then used to build a body of the next generation. But nothing remotely like that happens on our planet. And, between you and me, I suspect that it wouldn't ever work, not on any planet. One reason for this (only one reason out of several) is that a scan of the parent's body couldn't help reproducing things like scars and broken legs. Each generation would accumulate the scars and broken limbs of all the ancestors.

Yes, DNA is a digital code, just like computer code. And yes, DNA transmits digital information from parents to children and so on down countless generations. But no, the information transmitted is *not* a blueprint. It is not in any sense a map of a baby. It's not a scan of a parent's body. A genetics lab can read it, but it couldn't print out a baby. The only way to turn human DNA information into a baby is to put the DNA into a woman!

If DNA is not a blueprint of a baby, what is it? It's *a set of instructions for how to build a baby*, and that's a very different matter. It's more like a recipe for making a cake. Or like a computer program whose instructions are obeyed in order: first do this, then do that, then if so-and-so is true do . . . otherwise do . . . and so on for thousands of instructions. A computer program is like a very long recipe, complicated by branch points. A recipe is like a very short program, with only a dozen or so instructions. And a recipe is not reversible, like the building of a car or a house is. You can't take a cake and reconstruct the recipe by taking measurements. And you can't reconstruct a computer program by watching what it does.

The way houses are built is called 'top down'. In this sense of 'top', the architect's plan is at the top. The architect draws a set of detailed plans: a plan with precise dimensions for each room, detailed instructions as to what each wall is made of, how it is to be finished, where the water pipes and electric cables are to run, exactly where each door and window is to be, the precise location of every chimney and fireplace and supporting lintel. These plans are passed down to bricklayers and carpenters and plumbers, who take them and follow them meticulously. That's top-down building, with the architect – or rather, the architect's plans – directing the whole procedure from the top. That's 'blueprint building'.

Bottom-up building is very different. The best example I know is a termite mound. Look at plate 10 and be amazed. Daniel Dennett made a fascinating comparison to illustrate the distinction between bottom-up and top-down design – and the potential similarity, and complexity, of the results. On the right of this pair of illustrations is La Sagrada Família, a beautiful church in Barcelona. On the left is a termite mound, photographed by Fiona Stewart in Iron Range National Park in Australia. It's a mud nest built by a colony of termites. Actually, most of the nest is underground. The 'church' on the surface is an elaborate set of chimneys whose purpose is ventilation and air conditioning of the underground nest.

The resemblance is almost spooky. But the Barcelona church was *designed*, down to the last detail, using blueprints. Designed by the famous Catalan architect Antoni Gaudí (1852–1926). Nobody and nothing, not even DNA, designed the termite mound. Individual termite workers built it by following simple rules. No termite has the foggiest idea of what a termite mound should look like. None of them has anything like a picture or plan of a mud church in its brain or in its DNA. There never was a picture, or a blueprint, or a design for a termite mound, anywhere. Each individual termite just follows a set of simple rules, on its own, with no idea of what the other termites are doing, and no idea of what the finished building will be like.

I don't know exactly what those rules are, but this is the kind of thing I mean by a simple rule: 'If you come across a pointy cone of mud, stick another dollop of mud on it.' Social insects make use of chemicals – coded smells called pheromones – as an important system of communication. So the rules followed by individual worker termites when building a tower might depend on whether a particular piece of the edifice smells like 'this pheromone' or like 'that pheromone'. When 'design' emerges from the obeying of simple rules, where there is no overall plan in existence, anywhere, it is called 'bottom-up', as opposed to 'top-down', design.

Plate 11 shows another beautiful example of bottom-up 'design', starlings flocking in vast numbers in winter. In this case what's being 'designed' is behaviour, a sort of aerial ballet rather than a building. So, instead of saying 'There's no architect' I'm going to say 'There's no choreographer'. Nobody knows quite why they do it, but as evening approaches the birds congregate in huge flocks which can contain thousands of individuals. They fly together, fast and with such precise coordination that they don't collide, wheeling and turning together as though following direction from a master bird. A flock of starlings moves like a single animal. The 'animal' even has a distinct and definite edge. You should really look at some of the breathtaking movies of this wonder of the world. Search YouTube for 'Starling winter flocks'.

While you watch these flocks wheeling and soaring and diving, as though this huge conglomeration of birds were one giant animal, you can't help feeling there must be a master flight-coordinator, perhaps a single boss bird communicating with the others by telepathy: 'Turn left now, wheel up and around, now swing to the right . . .' It looks totally top-down. But it isn't. There's no director, no conductor, no architect, no boss bird. In a way now coming to be understood, all the individual birds, each one following bottom-up rules, together produce an effect which looks top-down. It's like the termites again, but on a faster time-scale. And what they produce is not a mud church but a superb aerial ballet – with no choreographer.

The power of this bottom-up non-choreography was beautifully demonstrated by a clever computer programmer called Craig Reynolds. He wrote a program called Boids to simulate flocking birds. You might think that Reynolds programmed the whole movement pattern of the entire flock. He didn't: that would be top-down programming. Instead, his bottom-up program worked like this. He put a lot of effort into programming just one bird, with rules such as: 'Keep an eye on your neighbouring birds. If a neighbour does so and so, you must do such and such.' Having perfected the rules for his one bird, he then 'cloned' it: made dozens of copies of the one bird and 'released' them all into the computer. Then he watched how the whole flock behaved. The boids flocked very much like real birds. Plate 12 shows a yet more beautiful simulation, building on Reynolds's one, programmed by Jill Fantauzza for the San Francisco Exploratorium.

The important point is that Reynolds didn't program at the flock level. He programmed at the level of the individual bird. Flock behaviour *emerged* as a consequence. Such 'bottom-up' programming is also how embryology works, with individual cells in an embryo playing the role of individual birds in a flock. Embryological development involves a lot of *movement* of cells, with membranes and sheets of tissue folding and caving in dynamically. So, as with the flying starlings, we are talking about 'no choreographer' as well as 'no architect'.

Embryologists work on how DNA builds a baby. Quite a lot is now known, but I'm not going to discuss it in detail. It would take a whole book, and it's not what this book is about. For our purposes, we just need to understand that embryonic development, the process by which bodies are built, is a bottom-up process. Like the way termite mounds are built, or flocks of starlings are coordinated. There is no blueprint. Instead, every cell in the developing embryo follows its own little local rules, like individual termites building a mud cathedral or individual starlings in a wheeling flock.

I'll go just a little bit further, into very early embryo life, to show how these bottom-up rules work. The fertilized egg, as you know, is a single cell. A big one. It splits into two. Then each of the two splits, to make four. Then those four split to make eight, and so on. After each split, the total size remains the same as the original fertilized egg. The same material is divided among two, four, eight, sixteen cells and so on, forming a solid ball. By the time the number of cells has reached a hundred or so, they have formed themselves (following local bottom-up rules) into a hollow ball, called the blastula. Once again, the size of the blastula is about the same as that of the original fertilized egg cell, and the cells themselves are now very small. The outside of the ball is a wall of cells.

The number of cells goes on increasing, as the cells split again and again. But the ball doesn't become bigger. Instead, again by each cell following local rules, part of the wall becomes dented in towards the middle of the ball. Eventually, the denting has gone so far that the ball is lined by two layers of cells instead of only one. The double-walled ball is called the gastrula, and the process of making it is called gastrulation.

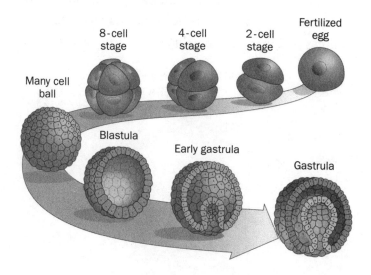

Admittedly a gastrula is not very complicated, and it doesn't look at all like a baby. But I think you can see how bottom-up rules followed by each cell, working on its own, could form the gastrula – by expanding the wall of the blastula and causing it to dent in to make the double-walled gastrula. And it's bottom-up rules like this that continue, working locally all over the embryo, to change the shape so that it steadily becomes more like a baby.

After gastrulation, another somewhat similar 'denting' process occurs. In this one, called 'neurulation', the denting ends up by pinching off a hollow tube, which is destined eventually to turn into the main nerve cord (the one that in each of us runs all the way down the back inside the spine). Again, the denting in neurulation works by individual cells following bottom-up local rules. The picture here shows how the nerve tube is made, first by 'denting' and then by a 'pinching off' of the dented part.

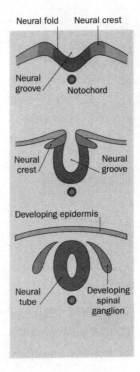

The details are different from gastrulation. But the same principle of bottom-up local rules is at work.

You remember how Craig Reynolds wrote a computer simulation of a flock of birds – 'Boids' – by programming the behaviour of just one 'boid'. He then made lots of copies of his one 'boid' and watched how they behaved together. They formed a flying, wheeling flock, just like real birds. Reynolds never programmed flock behaviour. Flock behaviour *emerged*, bottom-up, as a consequence of individual boids following local rules. Well, a mathematical biologist called George Oster did the same kind of thing, but with cells in an embryo instead of boids. He wrote a computer program to simulate the behaviour of a single cell. To do this he used lots of details that biologists already knew about single cells. Really quite complicated details, because cells are complicated things. But the important point is this. As with the boids, Oster didn't program an embryo. Just a single cell. Including the tendency to divide, which is one of the important things cells do. But cells do other things too, and Oster programmed them into his single cell, as well. He then let it divide on the computer screen, to see what happened.

As the cell divided, each copy inherited the same properties and the same behaviour as the original cell. So it was like Craig Reynolds cloning up lots of copies of his single boid, to see how they would behave in a flock. And, just as Reynolds's boids flocked like starlings, Oster's cells . . . well, just look at the diagram on this page to see what they did. And compare it with the picture of real neurulation, above. Of course, the two are not exactly the same. Nor were Reynolds's flocking boids exactly the same as real flocking starlings. In both cases, all I'm trying to do is show you the power of bottom-up 'design' where there is no architect/choreographer, only low-level local rules.

Later stages of embryology are too complicated to deal with here. Different tissues – muscle, bone, nerve, skin, liver, kidney – all grow by cell division. The cells of each tissue look very different from each other, but all have the same DNA. The reason they are different is that different stretches of DNA – different genes – are turned on. In any one tissue, only a small minority of the tens of thousands of genes are turned on. What this means is that in each tissue, the proteins, those vital 'lab-assistant' enzymes, that are made in the cells of that tissue are only a small minority of the enzymes that could be made – and actually are made in other tissues. And that leads to the cells in different tissues growing differently. Each tissue grows by cell division following local bottom-up rules. And each tissue stops growing when it reaches the right size: again, following bottom-up rules. Sometimes things go wrong and a tissue fails to stop growing: cells disobey the bottom-up rules that tell them to stop dividing. That's when we get a tumour, like a cancer. But mostly that doesn't happen.

Now let's put the idea of bottom-up embryology together with the crystals of Chapter 9. Crystals – pyrites or diamonds or snowflakes – grow their pretty shapes by local bottom-up rules. In those cases the rules are the rules of chemical bonds. We likened the molecules organized by those rules to soldiers on parade. The important point is that nobody designed the shape of the crystal. The shape emerged through the obeying of local rules.

Then we saw how the laws of chemical bonds – by a process that resembles jigsaw pieces slotting into each other – produced more elaborate things than ordinary crystals: protein molecules. Then the same kind of jigsawing caused the protein chains to coil up into 'knots'. And the 'crevices' in the 'knots' enabled them to act as enzymes, catalysts that turn on very particular chemical reactions inside cells. As I said before, 'crevices' is a great oversimplification. Some of these knotted molecules are tiny machines, miniature 'pumps', or tiny 'walkers' which literally stride about on two legs inside the cell, busily doing chemical errands! Look on YouTube for 'Your body's molecular machines' and be utterly amazed.

Enzymes switch on other enzymes which, in turn, catalyse other particular chemical reactions. And those chemical reactions inside cells cause the cells to work together, following local rules as in George Oster's simulation, to make an embryo. And then a baby. And every step of the way is controlled by DNA, again using just the same jigsaw rules. It's like crystals all the way through, but elaborate crystals of a very special kind.

The process doesn't stop with birth. It goes on as the baby grows into a child, the child grows into an adult, and the adult grows older. And of course, differences in DNA in different individuals – ultimately caused by random mutations – cause differences in the proteins that 'crystallize' or 'tie knots' under the influence of the DNA.

And the knock-on effects of those differences eventually show themselves, way down the line, in differences in the adult body. Perhaps the adult cheetah runs just a little bit faster. Or slower. Perhaps the chameleon's tongue shoots out just that little bit further. Perhaps the camel can cover just a few more miles of desert before dying of thirst. Perhaps the rose thorn is just a tiny bit sharper. Perhaps the cobra's venom is just a tad stronger. Any mutation in the DNA can have an effect, at the end of the long, long chain of intermediate effects on protein and cell chemistry and embryonic growth patterns. And that can make the animal more, or less, likely to survive. And that makes it more, or less, likely to reproduce. And that makes the DNA responsible for the change more, or less, likely to find itself in the next generation. So, as the generations go by, over thousands and millions of years, the genes that survive in the population are the 'good' genes. Good at building bodies that run fast. Or have long tongues. Or can go more miles without water.

That, in a nutshell, is Darwinian natural selection, the very reason why all animals and plants are so good at what they do. The details of what they are good at are different for each species. But it's all ultimately about being good at one thing: surviving long enough to pass on the DNA that makes them good at whatever it is they do. After thousands of generations of this natural selection, we notice (or we would if we lived long enough) that the average form of the animals in the population has changed. Evolution has occurred. After hundreds of millions of years, so much evolution has happened that an ancestor looking like a fish has given rise to a descendant looking like a shrew. And after billions of years, so much evolution has occurred that an ancestor like a bacterium has given rise to a descendant like you or me.

Everything about a living creature is the way it is because its ancestors evolved that way over many generations. That includes humans and it includes human brains. The tendency to be religious is a property of human brains, as is the tendency to like music and sex. It's therefore reasonable to guess that the tendency towards religious belief has an evolutionary explanation, like everything else about us. And the same goes for our tendencies, such as they are, to be moral, or to be nice. What might the evolutionary explanation be? That is the topic of the next chapter.

· 11 ·

Did we evolve to be religious?
Did we evolve to be nice?

Until pretty recently just about everybody believed in some sort of god. Outside western Europe, where only a minority nowadays are religious, most people around the world, including the United States, still do believe in a god or gods, especially if they aren't well educated in science. Shouldn't there be a Darwinian explanation of belief in gods? Did religious belief, belief in some kind of god or gods, help our ancestors to survive and pass on genes for religious belief?

I suspect that the answer is probably yes. Well, a kind of yes. Of course that doesn't mean that the gods people believe in – whichever gods those might be – are really there. That's a completely separate question. Believing in something that isn't really there could even save your life. There are various ways in which this might happen.

You remember the gazelles and zebras needing to strike a fine balance between being too scared and not scared enough? Now imagine you are an early human, long ago in our ancestral past on the African plains. Like a gazelle, you have to get the balance right between being sufficiently scared of lions and leopards, and being so scared that you never get on with the business of life. In the human case that might be the business of digging for yams or courting a mate. You hear a noise and look up from digging up a yam. You see a movement in the grass which just might be a lion. It could instead be the wind. You are making good progress in digging out a really big tuber and don't want to stop. But that noise just could be a lion.

If you believe it's a lion and it really is a lion, that valid belief might save your life. That's easy to understand. The next part is harder to understand. Even if it is *not* a lion on this particular occasion, a general *policy* of believing that mysterious movements or sounds spell danger could save your life. Because sometimes it really will be a lion. If you take that too far and run scared from every rustle in the grass, you'll miss out on the yams and the other business of living. But an individual who gets the balance right will still, on some occasions, find himself believing it's a lion when it actually isn't. And that tendency to believe what may turn out to be a falsehood will sometimes save your life. That's one way in which believing in things that don't exist could save your life.

Here's a slightly more technical way of putting the point. Humans have a tendency to believe in *agency*. What is agency? Well, an agent is a thing that deliberately does something for a purpose. When the wind rustles the long grass, there is no agency. Wind is not an agent. A lion is an agent. A lion is an agent whose purpose is to eat you. It will modify its behaviour in sophisticated ways in order to catch you, and work energetically and flexibly to thwart your efforts to escape. It's worth being scared of agency. But it can be a waste of time and effort, because the suspected agent may be something like the wind. The more dangerous your life tends to be on average, the more the balance should shift towards seeing agents everywhere and therefore sometimes believing falsehoods.

Nowadays we mostly no longer have to be scared of lions or sabretooths. But even modern humans can be scared of the dark. Children are scared of bogeymen. Adults are scared of muggers and burglars. Alone in bed at night, you hear a noise. It could be the wind. It could be the timbers of the old house, creaking as they settle. But it could be an armed burglar. Maybe nothing so specific as a burglar. As far as you are concerned, you fear an unnamed agent, as opposed to a non-agent like the wind or a creaking beam. The fear of agents, even if irrational, even if inappropriate on this particular occasion, may lurk within us from our ancestral past. My colleague Dr Andy Thomson put it like this in his book *Why We Believe in God(s)*: we are likely to mistake a shadow for a burglar; we are unlikely to mistake a burglar for a shadow. We have a bias towards seeing agents, even when there aren't any. And religion is all about seeing agency all around us.

Our ancestors' religions were 'animistic': they saw agents everywhere they looked, and often they called them gods. This is how the Greek gods started out, as is clear from Stephen Fry's lovely book *Mythos*. All over the world there were river gods and thunder gods, sea gods and moon gods, fire gods and sun gods, gods – or perhaps demons – of the dark forest. The sun was a god, an agent who had to be wooed and placated with prayers and sacrifices, otherwise he might decide not to rise tomorrow. The fire was a god who needed feeding or he'd go out. Thunder was a god – what else but a god could account for such a terrifying noise? The weather was so unpredictable, yet so important to life, it was natural to think agents were behind its changing moods. Surely there must be a way to end the terrible drought? A really big sacrifice to the rain god might do it. A terrible storm just wrecked our house. Perhaps we didn't heap sufficient praise on the storm gods and they were angry.

Yahweh evolved in people's minds to become the one God of the Jews, and eventually of Christians and Muslims too. Before that he was a 'storm god', one of many gods of the Canaanite peoples from whom the Jews sprang. Other Bronze Age Canaanite gods who were originally worshipped alongside Yahweh included Baal the fertility god and El, the chief god, and his wife the goddess Asherah. According to some scholars of religious history, Yahweh was later merged in people's minds with El and Asherah to eventually become the one and only God of the Jews. So Bronze Age animism came to be pared down to Iron Age monotheism. Later, Christianity and Islam adopted the God of the Jews. And later still the storm god of the Canaanites evolved further sophistication and became the hero of books on theology by learned professors at Oxford and Harvard.

I suggested that people made sacrifices to the weather gods in the hope of breaking a drought. But why would they think it might help? The human brain is a pattern-seeker. Natural selection has built into our brains a tendency to notice patterns such as sequences: what follows what. We notice that thunder follows lightning, rain follows after grey clouds gather, crops don't grow if there is no rain. But 'what follows what' is complicated. 'What follows what' turns out to mean not 'what *always* follows what' but 'what *sometimes* follows what'. Pregnancy follows sexual intercourse, but only sometimes.

Often we think we notice a pattern when there really isn't one. Often we fail to notice a pattern when there really is one. The mathematicians known as statisticians distinguish two ways of getting things wrong when we try to recognize these patterns. They call them *false positives* and *false negatives*. A false positive is thinking you see a pattern when there isn't one. Superstition is a common type of false positive error. A false negative is failing to notice a pattern when there really is one. There's a real pattern between being bitten by a mosquito and getting malaria. But it doesn't invariably follow, and nobody picked up on it until Sir Ronald Ross in 1897. There's no real pattern between a black cat crossing your path and subsequent misfortune. But many superstitious people have believed that particular false positive.

Last year we prayed to the rain gods and it then rained. Surely that pattern must have meant something?

No, it was meaningless. A false positive. It was going to rain anyway. But it's hard to shake off the superstition.

> The child was ill with a fever. We sacrificed a goat to the gods and the child got better. So we'd better sacrifice a goat the next time somebody gets a high fever.

The immune system often cures people of malaria anyway. But try telling that to a superstitious person who is convinced that sacrificing a goat did the trick.

Even if you notice an unvaryingly repeated pattern – something follows something else reliably, every single time – it doesn't prove that the earlier event caused the later one. The church clock in the village of Runton Acorn always strikes the hour shortly before the clock in the neighbouring village of Runton Parva. But does the Runton Acorn clock *cause* the Runton Parva clock to strike? Observation alone can't settle the question. Not even repeated observation. The only sure way to demonstrate cause is an *experiment*. You have to *manipulate* the situation. Climb up into the Runton Acorn tower and stop the clock. Does the Runton Parva clock then fail to strike? Then experimentally set the Runton Acorn clock ten minutes fast. Does the Runton Parva clock still strike just after it? Of course you have to repeat the experiment a respectable number of times to rule out chance – random luck.

It takes a sophisticated, perhaps rather nerdish mind to do proper experiments to test whether an apparent pattern is really there. You'd have to be very nerdish indeed to bother to do the church clocks experiment. And if the question is whether a noise really is a lion, the experimental approach could be fatal. No wonder our ancestors resorted to superstition instead.

A famous experimental psychologist called B. F. Skinner demonstrated superstition in pigeons. His pigeons 'noticed' patterns that were not really there: false positives. Each of eight pigeons was placed in a separate box called a 'Skinner box'. Each box had an electrically operated feeding apparatus which could deliver food to the hungry pigeons. Normally, Skinner boxes are wired up to deliver food only when the bird does something, like peck a switch in the wall of the box. But Skinner did something different for this particular experiment. He severed the connection between the feeding apparatus and the birds' behaviour. Nothing that the birds did had any effect on whether they got fed. Food was delivered into the box sporadically, regardless of what the bird did. Or, indeed, if it did nothing.

The result was fascinating. Six of the eight birds developed superstitious habits of various kinds. One bird walked round and round in an anti-clockwise direction, making two or three turns between rewards. We could say it had a superstitious belief that turning anti-clockwise caused the food to come. A second bird repeatedly thrust its head into one of the upper corners of the box. It 'thought' that that was what persuaded the feeding apparatus to deliver. Two other birds developed a 'pendulum' habit with the head. They thrust the head fast to left or right, then brought it back more slowly. Another bird's superstitious habit was to toss the head upwards, as though throwing some non-existent object up in the air. And the sixth bird directed pecking movements towards the floor, without ever hitting the floor.

Skinner called it superstitious behaviour and I think he was right to do so. What must have happened is this. A bird just happened to make a particular movement, say thrusting its head up into the corner, immediately before the feeding apparatus clunked into action. The bird 'thought' (not necessarily consciously) it was its head movement that had caused the food to arrive. So it did it again. And, as it happened, that was just the right time for the next food consignment to arrive. Each bird learned a different superstitious habit, repeating whatever it happened to do before food arrived by chance. And that, it seems likely, is how our ancestors developed the habit of, say, praying, or sacrificing a goat, to cure a child of a fever. The other resemblance between Skinner's pigeons and humans is that, in different parts of the world, local peoples develop different superstitious beliefs. Just like the six different pigeons in their 'local' Skinner boxes.

Gamblers, too, whether at the roulette wheel or a one-armed bandit, are rewarded at random, whatever they do. A gambler thinks he notices that he has more luck when he wears his 'lucky shirt'. Or he once prayed for luck and promptly won the jackpot. Just like Skinner's pigeons, he does it again. Never wins the jackpot again but can't rid himself of the habit of praying. You can't influence the probability that a slot machine will deliver the jackpot. Or that the ball on a roulette wheel will land where you want it to. Yet gamblers from Monte Carlo to Las Vegas are riddled with superstitious beliefs that they can.

Long ago, before computers had screens, they printed things out on a teleprinter instead. Once, while working in my university computer room, I watched a student who was desperately impatient for the computer to respond. He repeatedly rapped his knuckles on the teleprinter, although he must really have known it couldn't possibly persuade the computer to hurry up. Maybe he had once done it just before the computer happened to spew out its results anyway and never rid himself of the superstitious habit. Like Skinner's pigeons.

Let's suppose that in a time of drought our ancestors took it into their heads to sacrifice to the rain god. Every day. And eventually the rain came. Maybe a lot of sacrifices were necessary before the rain god – as they thought – was persuaded. The superstitious people never tried the experiment of *not* sacrificing to the rain god, to see if rain would come anyway. That's what a scientist would do. But our ancestors weren't scientists. And they didn't dare risk *not* sacrificing to the rain god.

Of course, I'm speculating. But I think it's plausible. It's exactly the kind of thing many tribespeople do, to this day. And Skinner's experiments were not speculation. They really happened. Nor is it speculation that human gamblers trust in lucky numbers, lucky charms and prayers. People tend to pray, or develop superstitious habits, whenever there's uncertainty as to what will happen (what we call 'chance' or 'luck') and we want a particular outcome. Superstition in itself probably didn't help our ancestors to survive. But a general tendency to look for patterns in the world – making an effort to notice when events tend to be followed by important other events – probably did. And superstition was a byproduct of this. As with the zebras balancing the risk of being eaten against the risk of not eating enough, human pattern-seekers had to strike a balance between two risks: the risk of noticing a pattern when there isn't one (superstitious false positive) and the risk of failing to notice a pattern when there is one (false negative). A tendency to notice patterns was favoured by natural selection. Superstition and religious belief were a byproduct of that tendency.

Now here's another line of thought. Our earliest human ancestors lived in a dangerous place, the African savanna. There were poisonous snakes, scorpions, spiders and centipedes underfoot. There were pythons and leopards lurking in the trees, lions behind bushes, crocodiles in the river. Adults knew of these dangers but children needed to be told. Parents would surely have warned their children, just as parents in modern cities warn their children to look left and right before crossing a road. Natural selection would have favoured parents who warned their children. And natural selection would have favoured genes that built into child brains a tendency to believe their parents.

That much is easy to understand. Now for the tricky part. If adults ever gave children bad advice alongside good advice, the child brain would have no way to distinguish bad advice from good. If the child brain were capable of making that distinction, the adult advice wouldn't be necessary anyway. The child would just *know*, for instance, that snakes are dangerous. The whole point is that if children already knew, parents wouldn't need to tell them. So if, for some reason, a parent were to give a child useless advice – like 'You have to pray five times a day' – the child would have no way of knowing that it was useless. Natural selection simply builds into the child brain the rule 'Believe whatever your parents tell you'. And that rule will come into force even when 'what your parents tell you' is actually silly or untrue. Or just based on a pigeon-like superstition.

But, you are probably asking, why should a parent tell a child something silly or untrue? Well, the parents themselves were once children. They were once given advice by their own parents. They too had no way of judging which advice was good and which advice was useless or bad. Advice, whether good or bad, gets passed on to the next generation. As for how it got started in the first place, pigeon-like superstition was probably part of the story. As the generations went by, the useless or superstitious advice got modified, amplified, by the same Chinese Whispers effect we saw at work in Chapters 2 and 3. In different parts of the world, different advice would get passed on. Which is exactly what we notice *has* happened, when we look around the world.

Of course some intelligent children, when they grow up, look at the evidence and succeed in breaking away from bad or useless advice from previous generations – grow out of it. Think about the title of this book. But that doesn't always happen, and I believe this partly explains how religions get started and why they persist. It's a sort of *byproduct* theory. Useless or superstitious beliefs, like the need to pray five times a day, or the need to sacrifice a goat to cure malaria, get passed on as a byproduct of sensible beliefs – or rather, as a byproduct of child brains being shaped by natural selection to believe parents, teachers, priests and other elders. And that is favoured by natural selection, because much of what elders tell children is sensible.

The byproduct theory is a truly Darwinian explanation for religious beliefs. True Darwinian explanations are all about genes becoming more numerous in a population. There are other kinds of explanation which look a bit like Darwinian explanations but aren't really. For instance, whole groups or nations might survive better because of their religion. And this means the religion itself survives. Suppose two nations have different religions. One has a warlike god, like Yahweh/Allah. Or like the warlike gods of the Vikings. The priests of such gods preach the virtues of courage in battle. They teach, perhaps, that a warrior who dies a martyr's death will go straight to a special martyrs' heaven. Or will go straight to Valhalla. They might even promise beautiful virgins in heaven to those men who die fighting for the tribal god (do you, like me, feel sorry for the poor virgins?). The other nation has a peaceful god or gods. Their priests don't advocate war. They don't preach heavenly bliss for those who die fighting. Maybe they don't preach any kind of heaven at all. All other things being equal, which nation will have the bravest warriors? Which nation is more likely to conquer the other? And therefore, which of the two religions is most likely to spread? The question answers itself. It is a matter of history that the spread of Islam, from Arabia throughout the Middle East and the Indian subcontinent, was due to military conquest. And the same goes for the spread of Christianity by the Spanish conquerors in South and Central America.

There are other possible ways in which religions might help nations or tribes, as well as in warfare. It's been suggested – I think quite plausibly – that a shared religion, and shared myths, rituals and traditions, helped societies to bond together and cooperate in ways that benefited everybody in them. It may seem daft to pray for rain, since modern science knows that praying for rain can't affect the weather. But what if coming together in a rhythmic rain dance helps promote solidarity and cooperation in the tribe? It's worth a thought, and respected colleagues have given it one.* Another possible non-Darwinian reason for the flourishing of religion is that kings and priests exploited the faith of their people as a means of dominating their societies. Yet another (and actually this one is close to being truly Darwinian) is the theory that ideas *themselves* – I've called them 'memes', to distinguish them from genes – including religious ideas, compete against rival memes in a gene-like way, to become more numerous in minds. There's no space here to explore these various theories; I just mention them to give you an idea of the kinds of debates being pursued. But now I need to move on.

In Chapter 6, I promised I'd return to the question of why natural selection favours niceness – at least, a limited form of niceness, which might serve as a kind of evolutionary basis for morality, a sense of what is good, and the desirability of doing good things. But I must say first that I think the changes in morality that I talked about in Chapter 6 are more important. Natural selection may build into our brains the basis for a limited amount of niceness. But it builds in the basis for nastiness too. As so often, there's a balance. What has happened in history is that the balance has shifted. In the nice direction, as we saw in Chapter 6.

* For example, Jonathan Haidt in *The Righteous Mind* (London, Penguin, 2012) and Yuval Noah Harari in *Sapiens* (London, Vintage, 2014).

So, what is the evolutionary basis for niceness? In Chapter 8 we saw that evolution is all about successful genes becoming more frequent in the gene pool (that's what successful *means*). Genes that equip individuals to run faster (though not so fast that their legs break like a racehorse's) become more numerous. Genes that make moths, lizards and frogs harder to see against tree bark become more numerous. Genes that make parents care for their children become more numerous, because copies of the very same genes survive in the bodies of the children cared for. So, being nice to your own children is a no-brainer, as far as natural selection is concerned.

But it isn't only your own children who contain copies of your genes. So do your grandchildren, nieces, nephews, sisters and brothers. The more distant the relationship, the lower the probability that a gene will be shared. A gene for saving the life of your child or your sister has a 50 per cent chance of being shared by the child or sister. A gene for saving the life of a nephew has a 25 per cent chance of being in the body of the nephew saved. A gene for saving the life of a first cousin has a 12.5 per cent chance of being shared by the cousin saved.*

---

* Those figures have to be understood properly. It's a little tricky. You may have read that most of our genes are shared by everybody anyway. That's true, and we also share a majority of our genes with chimpanzees and many other animals. The figures I have given for relatives like cousins refer to the probability of a gene being shared by a relative *over and above* a kind of 'baseline' probability that everybody in the population shares it.

So natural selection favours individuals who take slight risks to save the life of, or otherwise help, a first cousin. But it favours taking a greater risk to save the life of a niece. And an even greater risk to save the life of a sister or a son. Not just to save their life directly, either, but to help them in any way, like feeding them, or protecting them from predators or sheltering them from the weather.

Theoretically, natural selection favours feeding a brother as much as it favours feeding a son. But in practice there are more opportunities to usefully feed a son or a daughter than a brother or sister. This is why parental care is more common than sibling care. Sibling care really comes into its own in social insects like ants, bees, wasps and termites. Also certain birds like acorn woodpeckers in America, and mammals like naked mole rats in Africa.

Animals can't be expected to 'know' who their close relatives are. Natural selection of genes doesn't build into bird brains a rule like 'Feed your children'. Instead, the brain rule is more like 'Feed anything that opens its mouth and squawks inside your nest'. That's how cuckoos get away with laying their eggs in the nests of other birds. The baby cuckoo usually hatches first, and it throws out the eggs that were laid by the foster mother. The foster parent obeys the rule that its genes planted in its brain: 'Feed anything that opens its mouth and squawks in your nest.' That's exactly what the baby cuckoo does – and so it gets fed.

Our wild ancestors probably lived in small, roving bands like baboons. Later, in small villages. Both would have been equivalent to extended families. Almost everybody in the village or band would have been your uncle or your cousin or your niece. So a brain rule like 'Be nice to everyone' would have been equivalent to 'Be nice to your genetic relatives'. Most of us no longer live in small villages. It's no longer true that everyone you know is a cousin or a niece or other relative. But the rule 'Be nice to everyone' still lurks in our brains. This could be part of the Darwinian reason why we have a tendency to be friendly to others.

Unfortunately, there is a flip side to the coin. In the brains of our ancestors in their small bands or villages, the rule 'Be hostile to anyone you've never met before' would have been equivalent to 'Be hostile to anyone who is not a relative'. Or 'Be hostile to anyone who looks very different from you and the people you know'. Such brain rules could provide the biological origins of racial prejudice. Or of hostility to anybody perceived as 'other', like recent immigrants.

But unconscious rules of thumb aren't all the human brain has to offer. Unlike ants and acorn woodpeckers, humans have the brain power, especially aided by language, to actually know who is related to whom. The brain rule 'Be nice to everyone' could be superseded by a more specific brain rule: 'Be nice to individuals whom you actually know are your relatives.'

The !Kung peoples of the Kalahari desert are thought to be as close as any modern people to our ancestors. The light brown !Kung were in South Africa long before black invaders arrived from the North. They are hunters and gatherers who live in family groups. Each group claims ownership of a hunting territory. If a man strays into the territory of a rival group, he is in danger unless he can persuade the owners that he is related to somebody in their group. On one occasion, a man called Gao was caught in an area called Khadum, outside his home territory. The residents of Khadum were hostile. But Gao managed to persuade them that someone in Khadum had the same name as Gao's father. And it turned out that someone else in Khadum was also called Gao. This suggested that they shared relatives. The Khadum people then accepted Gao and gave him food.

The mountains in the centre of New Guinea were isolated from the rest of the world for thousands of years. In the 1930s, Australian and American explorers were amazed to discover about a million people, the New Guinea highlanders, who had never seen anyone from the outside world. The first encounters were pretty frightening for both sides. Archaeology suggests that the New Guinea highlanders had been there for about fifty thousand years. Some tribes were still hunters and gatherers like the !Kung. Other tribes had shifted to growing crops around nine thousand years ago, only a little later than agriculture began, independently, in the Middle East,

India, China and Central America. The New Guinea highlanders are divided into hundreds of tribes speaking mutually unintelligible languages. And they are hostile to members of other tribes. As with the !Kung, that even includes hostility to neighbouring bands belonging to the same tribe but different kin groups. In some areas, men who wander into territory belonging to a different kin group are in danger of being killed. They can be saved by a conversation in which they explore whether they have any cousins or other relatives in common. If they can identify a shared kinsman they may part amicably. If not, a fight, possibly to the death, is likely.

In addition to kinship, there is another way in which natural selection can favour niceness, one that might be more important than kinship. The theory here is called Reciprocal Altruism. If I do you a good turn today, you are likely to do me a good turn tomorrow. And vice versa. That's 'reciprocation'. And 'altruism' is another word for being nice. So 'reciprocal altruism' means being nice back to someone who is nice to you.

Reciprocal altruism doesn't need conscious awareness. Natural selection can favour genes that build brains that reciprocate, even though they don't realize it. A scientist called Gerald Wilkinson did a nice study of vampire bats. These bats feed on blood, the blood of larger animals such as cows. They roost in caves during the day, and come out by night to search for food. Victims are quite difficult to find, but if a bat succeeds in finding one there is plenty of blood. So much so, that the vampire gorges itself and flies home to its daytime cave with a surplus in its stomach. But a bat that fails to find a victim is in danger of starving to death. Small bats live much closer to the borderline of dangerous starvation than we do, and Wilkinson convincingly demonstrated this.

When the bats return to the cave after a night's hunting, some of them will be starving. Others will have a surplus. Starving bats beg from gorged bats, who vomit up some of the blood in their stomachs to feed the starving ones. The next day the roles may be reversed. The ones that had been lucky the previous night may now be starving, and vice versa. So theoretically, each individual bat can benefit from being generous after a good night's foraging, in the expectation of repayment after a bad night.

Now, Wilkinson did a clever experiment. He worked with captive bats, taken from two different caves. Bats from the same cave knew each other but didn't know those from the other cave. Wilkinson experimentally starved one bat at a time. Then he put it with other bats to see if they would feed it. Sometimes he put it with familiar 'friends'. Other times he put the experimental bat with strangers from a different cave. Consistently, the result tended to be the same: if they already knew the starved bat, yes, they'd feed it; if they didn't know it – if it came from the 'wrong' cave – they wouldn't. Of course, it could also be that bats from the same cave were genetically related. Later work by Wilkinson and a colleague showed that reciprocation – paying back good turns – is more important than kinship in this case.

Wilkinson's result probably makes total sense to you. Because you are human and that's how humans often behave. We have a strong sense of who has done us a good turn. And we know to whom we have done a favour. We expect to be paid back. We feel a sense of debt that needs repaying, and a sense of guilt if we fail to do so. And we feel resentment, feel let down, if somebody fails to repay a debt or a good turn.

Now think back to our distant ancestral past. Put yourself in the position of somebody living in one of those small villages or bands. Not only would you know everybody and remember debts and obligations between particular individuals. You'd also know that you are probably going to live in the same village for the rest of your life. Everyone in the village is a possible giver of favours for a long time into the future. The brain rule 'Be nice to everyone, at least at first or until you have good reason not to trust them' could well be built in by natural selection. You never know when you may need a good turn repaid to you. And it's plausible that our brains today have inherited the same brain rule from our ancestors. Even if we now live in big cities where we keep meeting people we are never going to meet again, we still have the brain rule to be nice to everybody unless there's a good reason not to.

The idea of reciprocation, of exchange of favours, is at the root of all trade. Nowadays, few of us grow our own food, weave our own clothes, propel ourselves from place to place with our own muscle power. Our food comes from farms which may be on the other side of the world. We buy the clothes we wear, get around in a car or on a bicycle which we haven't the faintest clue how to make. We board a train or plane which was made in a factory by hundreds of other humans, not one of whom probably knew how the whole thing was put together. What we offer in exchange for all these things is money. And we've earned that money by doing whatever it is we *can* do, writing books and giving lectures in my case, curing people in the case of a doctor, arguing in the case of a lawyer, fixing cars in the case of a garage mechanic.

Most of us would have a hard time surviving if we were transported back ten thousand years to the world of our ancestors. Back then, most people grew or found, dug up or hunted their food. In the Stone Age it's possible that every man made his own spear. But there would have been expert flint-knappers who made especially sharp spear points. At the same time there may have been expert hunters who could throw a spear hard and accurately, but were not skilled at making spears in the first place. What could be more natural than an exchange of favours? You make me a good sharp spear and I'll give you some of the meat that I catch with it.

Later, in the Bronze Age and then in the Iron Age, specialist smiths offered metal spears in exchange for meat. Specialist farmers offered crops to the smiths, in exchange for the digging tools that they needed to cultivate them. Later still, exchange became indirect. Instead of 'I'll give you food if you make me the tools to get the food', people gave money, or its equivalent, such as a written IOU as a token of a promise to repay the debt in the future.

Nowadays direct barter (swapping) which doesn't involve money is rare. It's even illegal in a lot of places because it can't be taxed. But our entire life is dominated by our dependence on other people with different skills. And the brain rule 'When in doubt, be nice' is still present in our brains. Along with other equally ancient accompanying brain rules such as 'Be prepared to be suspicious unless you have built up a relationship of trust'.

So there does indeed seem to be some Darwinian pressure to be nice, which could serve as the original basis for our sense of right and wrong. But I think it's swamped by later learned morals, such as we discussed in Chapter 6. And nothing in this chapter has changed the conclusion of Chapter 5: we don't need God to be good.

Taking courage from science

Before Darwin came along, it seemed absurd to almost everyone that the beauty and complexity of the living world could have come into being without a designer. It required courage to contemplate even the possibility. Darwin had that courage, and we now know he was right. There are still unsolved problems in science – gaps in what we so far understand. And some people are tempted to say the same kinds of thing that were said about life before Darwin came along. 'We don't yet understand how the evolutionary process began in the first place, so God must have started it.' 'Nobody knows how the universe began, so God must have made it.' 'We don't know where the laws of physics come from, so God must have made them up.' Wherever there is a gap in our understanding, people try to plug the gap with God. But the trouble with gaps is that science has the annoying habit of coming along and filling them. Darwin filled the biggest gap of all. And we should have the courage to expect that science will eventually fill the gaps that remain. That is the theme of this final chapter.

It used to be simple common sense that living things had to be created by God. Darwin exploded that particular piece of common sense. This chapter sets out to undermine our confidence in common sense, beginning with relatively trivial examples and moving on to more important ones. Each example concludes with the refrain 'You cannot be serious!' (it's a memorable quote from the great tennis player John McEnroe, who frequently used it to query dubious line decisions). We then return to the bigger example: the apparent common sense that says there must be a God to explain the universe's origin and other so-far unsolved problems.

In 2014, a teenager was caught on camera urinating into a reservoir in America. The local water authority therefore took the decision to drain the reservoir and clean it at an estimated cost of $36,000. The volume of water drained was about 140 million litres. The volume of urine was perhaps about a tenth of a litre. So the ratio of urine to water in the reservoir was less than one part in a billion. There were dead birds and debris in the reservoir, and presumably plenty of animals had urinated into it without anyone noticing. But such was the 'yuck' reaction many people felt, the fact that a single human was *known* to have peed in the reservoir was enough to get it drained and cleaned. Is that sensible? What would you have done if you'd been in charge of the reservoir?

Every time you drink a glass of water, there's a high chance you'll drink at least one molecule that passed through the bladder of Julius Caesar.

You cannot be serious! But it's true.

Here's the reasoning. All the water in the world is continuously being recycled by evaporation, rain, rivers and so on. Most of it is in the sea at any one time, and all the rest of the world's water gets circulated through the sea as the decades go by. The number of water molecules in a glassful is about 10 trillion trillion. The total volume of water on the planet is about 1.4 billion cubic kilometres, and that corresponds to only about 4 trillion glassfuls. I say 'only' because 4 trillion is a tiny number compared to the 10 trillion trillion molecules in a glassful. So there are trillions of times more molecules in each glassful than there are glassfuls in the world.

Which is why it's safe to say you've drunk some of Julius Caesar's pee. Of course, there's nothing special about Julius Caesar. You could say the same of his friend Cleopatra. Or Jesus. Or anybody, provided there's been enough time for recycling to have taken place. And what's true of a glassful is true many times over of a reservoir. That American reservoir didn't only contain the urine of the teenager who was caught peeing in it. It contained the urine of millions of people, including Attila the Hun and William the Conqueror and very possibly you too.

Air is recycled in the same kind of way as water, only faster, and the same kind of calculation works here too. The number of molecules of air in a lung is hugely greater than the number of lungs in the world. You have almost certainly breathed in atoms that were breathed out by Adolf Hitler. And Hitler's secretary reported that he had bad breath.

Science can be very surprising. We're talking about the courage you need in order to cope with the surprises. Courage that should be applied to the mysteries that remain unsolved.

T. H. Huxley (Darwin's friend, whom we met in Chapter 1) said: 'Science is nothing but trained and organized common sense.' But I'm not sure he was right. The stories I'm telling in this chapter seem to defy common sense. Galileo defied common sense when he showed that, air resistance apart (you have to do the experiment in a vacuum), a cannonball and a feather, when dropped from a height, will hit the ground at the same moment.

You cannot be serious, Galileo! But it's true.

Here's why Galileo was right. According to Isaac Newton, every object in the universe is attracted to every other object by gravity. The force of the attraction is proportional to the masses of the two objects (think of mass, for the moment, as rather like weight – there is a difference, but we'll come to that in a moment) multiplied together. The cannonball is much more massive than the feather, so gravity will exert a stronger force on it. But the cannonball needs more force than the feather to accelerate it to the same velocity. The two exactly cancel out, with the result that feather and cannonball hit the ground together.

I said I'd clarify why mass is not the same as weight. On our planet, the mass of an object, such as a man, is the same as his weight, say 75 kilograms. But in the space station, the man is weightless. His weight is zero, while his mass is still 75 kilograms. A cannonball in the space station would float like a balloon. But you'd know it had plenty of mass if you tried to throw it across the cabin. It would need a lot of effort. As you shoved it, unless you were supported by a wall, you'd simultaneously shove yourself in the opposite direction. Not at all like a balloon. And when the cannonball hit the wall on the other side of the cabin it would crash in a 'massive', clunking way and might break something. If it hit somebody on the head it would hurt them (again, not like a balloon), even though both the cannonball and the head are weightless. The weight of a cannonball is a measure of the downward pull of Earth's gravity on the ball. Its mass is a measure of the total amount of matter it contains. If you were to weigh the cannonball in the space station, the weighing machine and the cannonball would both float around freely, so the cannonball would not exert any pressure on the weighing machine. It would have a weight of zero.

Same thing if you were to jump out of a plane, sitting on a weighing machine. Both you and the weighing machine would fall at the same rate. So again, you wouldn't press down on the weighing machine and it would report your weight as zero. Your weight is zero while you fall. But your mass remains, in full.

That gives you the clue as to why cannonballs (and men and weighing machines) float around weightless in the space station. Many people think it's because they are a long way from Earth and therefore beyond the pull of Earth's gravity. That is utterly wrong. It's a very common mistake. Actually, the pull of Earth's gravity is nearly as strong in the space station as it is at sea level, because the space station is not so very far away. The reason objects in the space station are weightless is that, like the person who has jumped out of a plane sitting on the weighing machine, they are continuously *falling*. Falling, in this case, *around* the Earth. The moon, too, is continuously falling around the Earth. The moon is weightless, although it has a mass of 10 thousand billion billion kilograms.

'The moon is weightless and continuously falling around the Earth?'

You cannot be serious! But it's true.

We think of our planet as all rough and wrinkly, pitted and studded with valleys and mountain ranges. After all, Mount Everest is nearly 9 kilometres high and the first two men to climb it were hailed as heroes for their feat. But if you were to shrink the Earth to the size of a ping-pong ball, the surface would feel smooth all over. Even

Everest wouldn't register to the touch: it would be as small as a grain of sand on the finest sandpaper.

You cannot be serious! But it's true.

Work it out for yourself. Measure a ping-pong ball; you know the height of Everest; look up the diameter of Earth and do the sum.

Why are planets round? Gravity pulls them inwards from all directions. Even solid ground behaves like a liquid, given enough time. Smaller objects like comets are not round but knobbly and misshapen. This is because their gravity is too weak to pull them into shape. Pluto is big enough to be spherical. However, it is smaller than several known 'planetesimals', which is why Pluto has been demoted from planetary status. This upset a lot of people. But it is only a matter of definition: a matter of 'semantics'. Mars, being smaller than Earth, has weaker gravity, and so less force to pull its mountains inwards. This is why Mars can (and does) have higher mountains than Everest. Mars as a ping-pong ball would feel infinitesimally rougher to the touch than Earth. But its tiny moons Phobos and Deimos are positively knobbly by comparison. They look like potatoes.

Once upon a time, it seemed obvious common sense that the world stood still and the sun, moon and stars revolved around it. What could be more natural? The ground you stand on feels rock steady. The sun moves across the sky from east to west daily, and so do the stars if you have the patience to note their changing positions. The Greek mathematician Aristarchus (*c.*310–230 BC) seems to have been the first to realize that the Earth orbits the sun. It's the Earth's spinning that makes it look as though the sun travels across the sky. This daring truth was forgotten for centuries until it was rediscovered by Nicolaus Copernicus in Poland (1473–1543). So contrary was it to common sense that Galileo was threatened with torture for promoting it.

You cannot be serious, Galileo! And we are going to torture you unless you recant.

If you look at a map of the world, you'll notice that the west coast of Africa and the east coast of South America look as though they might fit each other like jigsaw pieces. In 1912 a German scientist called Alfred Wegener had the courage to take this observation seriously and see where it led. He proposed that the map of the world changes. In a big way. Africa and South America, he suggested, really were once joined. He was ridiculed in his own lifetime. How could anything so massive as a continent split down the middle and the two halves – South America and Africa – drift thousands of miles apart? Yet that is what happened.

You cannot be serious! But it's true.

Wegener was right. Sort of. Until about 130 million years ago, Africa and South America really were joined.

Then they were slowly wrenched apart. There was a time when you could jump across the narrowest gap. A bit later you could swim across. Now the journey takes hours, even in a fast airliner. Wegener got the details a bit wrong. The evidence is now overwhelming that the entire surface of the Earth consists of interlocking and overlapping 'plates'. Like armour plates. They're called 'tectonic plates' and they move, but much too slowly for us to notice in our short lifetimes. Their movement has been compared to the rate at which fingernails grow. It's not smooth, however, like the growth of fingernails. More jerky, with no perceptible movement for a while, then a sudden movement like an earthquake. Indeed, it often is an earthquake.

Tectonic plates don't just consist of land. Much of each plate is under the sea. The continents are just the highlands riding on top of the plates. It's the plates that move, carrying the continents on top of them. There are no gaps between the plates. Where they push against each other, various things can happen. Including earthquakes. The two plates may slide past each other (that's what's happening at the famously earthquake-ridden San Andreas fault in western North America). Or one may slide under the other. This 'subduction' can push up a great mountain range like the Andes. Or the Himalayas, which were raised when the plate carrying India, then a huge island travelling north, forced its way under the Asian plate. The evidence for plate tectonics is fascinating and totally convincing. But I won't go into it here, because I did so in *The Magic of Reality*. I'll just point out that it's highly surprising and violently opposed to common sense.

Now for something so surprising it's actually frightening. At least, I think it is. You, and the chair you sit on (the table you eat off, the solid rock you stubbed your toe on) consist almost entirely of empty space.

You cannot be serious! But it's true.

All matter consists of atoms, and every atom consists of a tiny nucleus orbited (for want of a better word, although it's a bit misleading) by a cloud of far tinier electrons. Between them – nothing but empty space. Diamonds are proverbially hard. As we saw in Chapter 9, a diamond is a crystal lattice made of precisely spaced carbon atoms. If you imagine a carbon nucleus swollen to the size of a tennis ball, the nearest neighbour tennis ball in a diamond lattice would be 2 kilometres away. And the space between them would be empty because electrons are too small to matter. If you could shrink yourself to a scale where you could hit one of those balls with your tiny racquet, the next nearest tennis balls in the lattice would be much too far away for you to see.

My colleague Steve Grand, in his book *Creation*, wrote:

> Think of an experience from your childhood. Something you remember clearly, something you can see, feel, maybe even smell, as if you were really there.

After all, you really were there at the time, weren't you? How else would you remember it? But here is the bombshell: you weren't there. Not a single atom that is in your body today was there when that event took place . . .

You cannot be serious! But it's true.

Matter flows from place to place and momentarily comes together to be you. Whatever you are, therefore, you are not the stuff of which you are made. If that doesn't make the hair stand up on the back of your neck, read it again until it does, because it is important.

Does that mean that a man who has just been arrested for a crime he committed 30 years ago cannot be guilty because he's no longer the same person? What would you say if you were on a jury and the defence lawyer made that argument?

Here's something else that's pretty alarming. It follows from Albert Einstein's Special Theory of Relativity. If you set off in a spaceship at nearly the speed of light, and came back after your onboard calendar told you you'd been away 12 months, you would have aged only one year while all your friends back on Earth had died of old age. The world would be hundreds of years older, but you would be only one year older. Time itself on the spaceship, including all clocks and calendars on board, as well as the ageing process, would slow down as far as people on Earth were concerned. But not as far as everyone in the spaceship was concerned. On board the spaceship, everything would seem completely normal. So, back on Earth, your own great-great-grandson could be older than you, with a long white beard.

You cannot be serious! But it's true.

The message of this chapter is that science regularly upsets common sense. It serves up surprises which can be perplexing or even shocking; and we need a kind of courage to follow reason where it leads, even if where it leads is very surprising indeed. The truth can be more than surprising, it can even be frightening. I myself find the sheer weirdness of quantum theory positively frightening. Yet it must in some sense be true, because experiments have verified the mathematical predictions of quantum theory to an accuracy equivalent to predicting the width of North America to within one hairsbreadth.

What is the 'weirdness' that I'm talking about? There's no space here to go into all the shatteringly strange experimental results. I will just mention the so-called 'Copenhagen Interpretation' of some of these weird experimental results. The Copenhagen Interpretation says that some events, quantum events, haven't happened until somebody looks to see whether they have happened. It sounds daft, and the idea was satirized by the Austrian physicist Erwin Schrödinger, one of quantum theory's founding fathers. Schrödinger imagined a cat shut up in a box in which there is a killing mechanism triggered by the kind of event which is called a quantum event. Until we open the box, we don't know whether the cat is dead or not. But surely it must definitely be *either* alive *or* dead? Mustn't it? Not according to the Copenhagen Interpretation. According to the Copenhagen Interpretation, as satirized by Schrödinger, the cat is neither alive nor dead until we open the box to have a look. Obviously absurd, and that was Schrödinger's point. Yet, however absurd, it seems to follow from the Copenhagen Interpretation. And the Copenhagen Interpretation is favoured by many distinguished physicists. Somebody just sent me a lovely cartoon. The scene is a veterinary waiting room with pet owners patiently waiting. The nurse comes out and speaks to one of the gentlemen: 'About your cat, Mr Schrödinger. I have some good news and some bad news.' Now that's witty.

The apparent absurdity of the Copenhagen Interpretation has driven other physicists to an alternative interpretation called the Many Worlds Interpretation of quantum theory (not to be confused – though it often is – with the Multiverse Theory, which I'll come on to in a moment). According to the Many Worlds Interpretation, the world is continuously splitting into trillions of alternative worlds. In some of those worlds the cat is already dead. In other worlds the cat is alive. In some of those worlds I am already dead. In other worlds (necessarily including the world in which I am typing these words) I am still alive. In yet other worlds (not many) I have a green moustache. The Many Worlds Interpretation seems in one way less absurd than the Copenhagen Interpretation. In another way more so. Don't worry if you are totally bewildered by this paragraph and the previous one. So am I. That is precisely the point I am making. Scientific truth is frightening and we need courage to face up to it.

In an earlier century, Galileo's persecutors were frightened by the heretical idea that the Earth spins, and moves around the sun. Anyone might be frightened when they first discover that they and the solid earth they stand on are almost entirely empty space. But that doesn't stop it being true. And far more often than it is bewildering or frightening, scientific truth is wonderful, beautiful. You need courage to face the frightening, bewildering conclusions of science; and with the courage comes the opportunity to experience all that wonder and beauty. The courage to cut yourself adrift from comforting, tame apparent certainties and embrace the wild truth. Like my friend Julia did when she lost her Christian faith.

Julia Sweeney is an American comedian and actor. She wrote and performed a charmingly comic stage show called *Letting Go of God*. Julia was a good Catholic girl. When she grew up she started to question her faith. She thought hard and long about it. Lots of things didn't make sense. Many aspects of her religion seemed bad to her, rather than good as she had been taught. She read books on science and books on atheism. Then one day, when her habit of questioning had reached an advanced stage, she heard a little voice in her head. At first it was no more than a whisper: 'There is no God.' It grew louder: 'There is no God.' Finally, a panicked heart cry: 'OH MY GOD, THERE IS NO GOD!'

> I sat down and thought, 'Okay. I admit it. I do not believe there is enough evidence to continue to believe in God. The world behaves exactly as you would expect it would, if there were no supreme being, no supreme consciousness, and no supernatural.
>
> And my best judgment tells me that it's much more likely that we invented God than that God invented us. And I shuddered. I felt I was slipping off the raft . . .
>
> But then I thought, 'But I don't know how to not believe in God. I don't know how you do it. How do you get up, how do you get through the day?' I felt unbalanced. I thought, 'Okay, calm down. Let's just try on the not-believing-in-God glasses for a moment, just for a second. Just put on the no-God glasses and take a quick look around and then immediately throw them off.' And I put them on and I looked around.
>
> I'm embarrassed to report that I initially felt dizzy. I actually had the thought, 'Well, how does the Earth stay up in the sky? You mean, we're just hurtling through space? That's so vulnerable!' I wanted to run out and catch the Earth as it fell out of space into my hands.
>
> And then I remembered, 'Oh yeah, gravity and angular momentum is gonna keep us revolving around the sun for probably a long, long time.'

Julia bravely followed evidence and reason, even though this led her out of her childhood comfort zone.

This chapter is about the steps of courage that you need to take on the road to atheism. A pretty big step concerns the origin of the entire universe. We'll come to that later. But, as I said in my introduction to this chapter, an even bigger step was to understand the evolution of life. And that's a step humanity has already taken. We should take courage from that.

I've often wondered why it took until the middle of the nineteenth century for humanity – in the shape of Charles Darwin – to tumble to the full truth of evolution. Evolution by natural selection, as I hope Chapters 8 and 9 have demonstrated, really isn't very difficult to understand. You don't need mathematics to get the principle. Darwin was no mathematician; nor was Alfred Wallace, who discovered the idea independently, and only slightly later. Why did nobody get it before the nineteenth century?

Why didn't Aristotle (383–322 BC) get it? He is regarded as one of the world's great thinkers. He pretty much invented the principles of logical thought. He observed and described animals and plants in meticulous detail. Yet he was totally clueless when it came to answering the obvious question they raise, namely 'Why are they there?' Archimedes (c.287–212 BC) had some supremely clever ideas, both in and out of his bath (search it on the web, although unfortunately the story of Archimedes leaping out of his bath may be another of those repeat-worthy myths like the ones we met in Chapter 3). But the idea of evolution by natural selection never occurred to him. Eratosthenes (276–194 BC) calculated the circumference of the Earth by comparing the length of a midday shadow at two places a known distance apart. Brilliant! He accurately estimated the tilt of the Earth's axis (the tilt that gives us our seasons). Those feats are far cleverer than anything most of us could aspire to. Yet, although those clever old Greeks were surrounded by animals and plants (and of course humans), and they must have wondered how they came to be so purposeful, so beautifully 'designed', they never hit upon the extremely simple idea – Darwin's idea. Nor did Galileo. Nor did Isaac Newton, who just might be the cleverest person who ever lived.* Nor did any of the great philosophers throughout history. The idea is so simple and so powerful, you'd think any fool could have seen it; any fool sitting in an armchair, with no great learning and no mathematics. You'd think it would be

---

* Newton was a complicated mixture of contradictions. A superbly rational scientist, he wasted much of his life on a fool's errand to change base metals into gold. And much of the rest of his life on other fool's errands such as analysing the Bible for the significance of numbers mentioned there. By the way – not that it has any bearing on his cleverness – he wasn't a very nice man, unlike Darwin. He treated his rival Robert Hooke badly, although the jealousy, you might think, should have run the other way. As against that, when his dog Diamond upset a lamp and burned some important papers Newton had been working on, he didn't lose his temper but simply exclaimed, 'Oh, Diamond, Diamond, thou little knowest the mischief thou hast done!' That, at least, is how a well-known story goes. Some historians claim it never happened. In which case it's yet another good example, to add to those of Chapter 3, of how myths get started.

easier to solve than an average crossword clue (I speak with feeling, as I am hopeless at cryptic crosswords). Yet nobody hit upon it until the middle of the nineteenth century. This breathtakingly powerful yet simple idea, which had eluded the world's greatest minds, finally occurred to two non-mathematical travelling naturalists and specimen-collectors, Charles Darwin and Alfred Wallace. It also seems to have occurred independently, around the same time, to a third man, a Scottish orchard-keeper called Patrick Matthew.

Why did it take so long? Here's what I think. I think the complexity, beauty and 'purposefulness' of living things must have seemed too *obviously* designed by an intelligent creator. So it required a major leap of courage to consider anything else. I don't mean physical courage, like the courage of a soldier in battle. I mean intellectual courage: the courage to contemplate the apparently ridiculous, and say: '*You cannot be serious – but let's in any case go out on a limb and examine the possibility all the same.*' It had been 'obviously' ridiculous to suggest that a cannonball and a feather will fall at the same rate. But Galileo had the intellectual courage to examine the possibility and prove it. It seemed completely ridiculous that Africa and South America were once united and slowly drifted apart. But Wegener had the courage to see where the idea led. It must have seemed utterly ridiculous that something as obviously 'designed' as a human eye is actually not designed at all. But Darwin had the courage to examine that 'ridiculous' possibility. And now we know he was right. Right about that, and right about every last detail of every living thing.

The simple truth of evolution by natural selection was staring all those clever Greeks, all those brilliant mathematicians and philosophers before Darwin, in the face. But none of them had the intellectual courage to defy what seemed obvious. They overlooked the wonderful bottom-up explanation for what seemed, wrongly, to have top-down creation written all over it. The fact that the true explanation is so blindingly simple meant that it took even more courage to pursue it and work it out in detail. Natural selection evaded all those brilliant minds precisely because it is so simple. Too simple, one might have thought, to do the heavy lifting of explaining the whole of life in all its complexity and diversity.

We now know – the evidence brooks no alternative – that Darwin was right. There are a few details left to clean up. For example, we still don't know – yet – exactly how the process of evolution got started, some four billion years ago. But the main mystery of life – how did it come to be so complex, so diverse and so beautifully 'designed' – is solved. And my final point in this book is that Darwin's and Galileo's and Wegener's intellectual courage should inspire us to go further, in the future. All those examples of apparently ridiculous propositions turning out to be true should give us new courage when we face the remaining big puzzles of existence. How did the universe itself begin? And where do the laws that govern it come from?

A word of caution, by the way, before we go on. Galileo, Darwin and Wegener proposed daringly surprising ideas and they were right. Plenty of people propose daringly surprising ideas and are wrong, crazy wrong. Courage isn't enough. You have to go on and prove your idea right.

Our view of the universe has swelled over the centuries. And the universe itself is literally swelling as the seconds tick by. Once, people thought the Earth was pretty much all there was, with the sun and moon circling overhead, and the stars little peepholes through a hemispherical shell into heaven. Now we know the universe is large beyond all contemplation. But we also know that, once upon a time, the universe was small beyond all contemplation. And we know when that was. It was, according to current estimates, about 13.8 billion years ago.

The expanding universe was a twentieth-century discovery. There are people alive in the world today – my 102-year-old mother is one – who were born into a universe consisting of a single galaxy. Now she lives in a universe of 100 billion galaxies rushing away from each other as space itself expands. That's not an accurate way to put it, of course. She and Shakespeare and Galileo and Archimedes and the dinosaurs were all born into the same expanding universe. But when my mother was born, in 1916, nobody knew about anything other than the one galaxy we call the Milky Way. That was the universe. In

Galileo's time nobody even knew about that. Scientific truths are true even if there's nobody around to know about them; were true before humans appeared; will be true after we are extinct. That's an important point that evades many otherwise clever thinkers.

It is likely that even our expanding universe of 100 billion galaxies is not the only universe. Many scientists think – with good reason – that there are billions of universes like ours. On this view, ours is just one universe in a *multiverse* of billions of universes. We'll return to that idea in a moment.

Physicists today have a pretty good idea of what happened in the very early history of our universe. By 'very early' I mean back in the first tiny fraction of a second after the birth of the universe. And not only after the birth of the universe: after the birth of time itself. What can that possibly mean: 'the birth of time'? What happened before that? Physicists tell us we are not allowed to ask that question. It's like (or so they say) asking what is north of the North Pole. But that prohibition may apply only to our universe. That's if our universe is indeed one of billions in a multiverse.

God-worshippers nowadays (the educated ones, anyway) have given up on the living world as evidence of a creator. This is because they now understand that, where life is concerned, Darwinian evolution provides a full explanation. They've switched, instead, to other kinds of argument. With some desperation – or so it seems to me – they have turned their attention to other 'gaps'. Especially cosmology and the origin of everything, including the fundamental laws and constants of physics.

I need to explain what's meant by the fundamental constants of physics. There are some numbers that you can measure. Like the number of protons in a silver atom. There are other numbers that you can estimate. Like the number of water molecules in a glassful. And there are other numbers whose value is mathematically necessary. Like π (pi), the ratio of the circumference of any circle to its diameter – and π enters into mathematics in many other fascinating ways. But there are some numbers that physicists just accept without knowing why they have the values that they do. These are called the fundamental constants of physics.

An example is the gravitational constant, symbolized by the letter G. You'll remember we learned from Newton that all objects in the universe, such as planets, cannon-balls and feathers, are attracted to each other by gravity. The more distant the objects are from each other, the weaker the attraction is (it's inversely proportional to the distance multiplied by itself). And the more massive the two objects, the stronger the attraction between them (it's proportional to the two masses multiplied together). But to get the actual force of attraction itself you finally have to multiply by another number, G, the gravitational constant. Physicists believe G is the same all over the universe, but they don't know why it has the value it does.

It's possible to imagine an alternative universe with a different value of G. And if G were even slightly different, the universe would be very, very different.

If G were smaller than it is, gravity would have been too weak to gather matter into clumps. There'd be no galaxies, no stars, no chemistry, no planets, no evolution and no life. If G had been just a little bit bigger than it is, stars couldn't exist as we know them and they wouldn't behave as they do. They'd all collapse under their own gravity and perhaps become black holes. No stars, no planets, no evolution, no life.

G is just one of the physical constants. Others include c, the speed of light; and the 'strong force', which binds the atomic nucleus together. There are more than a dozen of these constants. Each of them has a value which is known but which is not (so far) explained. And in all cases you can say that, if their value were different, the universe as we know it could not exist.

This has led some theists to hope that God must be lurking somewhere behind the scenes. It's as if the value of each fundamental constant had been set by a knob that you might twiddle, like the tuning knob on an old-fashioned radio set. All the knobs had to be correctly tuned in order for the universe as we know it to exist – and therefore for us to exist. The temptation is to think that a creative intelligence – a god of some kind, a divine knob-twiddler – did the fine tuning.

It's a temptation that should be sternly resisted. For

reasons we've seen in earlier chapters. The fine-tuning of all those knobs might seem improbable, because there are so many other positions each tuning knob could be in. But, however improbable that fine-tuned precision may seem, any god capable of doing the precision tuning must be at least as improbable. How else would he know how to tune them? Importing a god into the reasoning doesn't solve the problem. It simply pushes it one stage back. It's a crashingly obvious non-explanation.

The problem Darwin solved, namely the problem of the massive improbability of life, was the big one. Before Darwin came along, the recurring phrase from the first part of this chapter, 'You cannot be serious!', would have hit home with immense force for anyone daring to question the divine creation of life. Perhaps with more force than for any other case. All that complexity, the speed and grace of a swallow, the fine-tuned flight surfaces of an albatross or a vulture, the bewildering intricacy of a brain or a retina, to say nothing of every one of the quadrillion cells of an elephant, the shimmering beauty of a peacock or a hummingbird – all that came about through the unaided, undirected, unsupervised laws of physics?

Explaining something so comparatively simple as the origin of the laws and constants of physics themselves should be a doddle by comparison. Admittedly we haven't solved that problem yet. But the success of Darwin and his successors in solving the bigger problem of life, and its fine-tuning to the needs of survival, should give us courage. Especially when we add to Darwin all the other spectacular successes of science. We're familiar with the list of such successes. Without antibiotics, vaccination and scientific surgery, many of us would be dead. Without scientific engineering, few of us would have travelled more than a few miles from where we were born. Without scientific agriculture, most of us would starve. Here, though, I want to pick out and focus on just one spectacular piece of science, one connected with the deep question we are concerned with – how did the universe come to be the way it is?

Cosmologists, working around the world and feeding constructively on each other's findings, have built up a detailed theory of what happened after the Big Bang. But how would you test such a theory? You'd need to set up 'initial conditions' – meaning the way you think things were immediately after the Big Bang. Then use the theory to deduce how things ought to be today, if your theory is correct. In other words, use your theory to predict the present from the distant past. Then look at the way things actually are to see if your prediction was correct.

You might think you could use mathematical proofs to deduce your prediction. Unfortunately the details are far too complicated for that. On top of the gravitational forces, there are billions and billions of tiny local interactions, for example in the swirling clouds of gas and dust. Such complexity can only be handled by constructing a 'model' in a computer and seeing what happens when you run it. Rather like Craig Reynolds and his 'Boids' model, which we looked at in Chapter 10. But much more elaborate. And when I said 'a computer', that's just short-hand: a single computer, however big, is nowhere near big enough to simulate the growth of the universe, the calculation is so huge. The most advanced simulation so far is called Illustris, and it needed not one computer but 8,192 computer processors running in parallel. And they weren't just ordinary computers, they were super-computers. The Illustris simulation begins not at the Big Bang itself but three hundred thousand years later (a very short time compared to the subsequent 13.8 billion years). Even all those supercomputers couldn't simulate every last detail of every atom. But it's fascinating, nevertheless, to compare the predicted shape of today's universe with the actual reality.

Take a look at plate 13, which contains a sort of joke. There is a top–bottom split in the picture. One half is the real universe, the famous Hubble Deep Field photo-graph, taken by the Hubble orbiting telescope in 1995. The other half is the universe as predicted by Illustris. Can you tell which is which? I can't.

Isn't science wonderful? If you think you've found a gap in our understanding, which you hope might be filled by God, my advice is: 'Look back through history and never bet against science.'

The Illustris simulation, as I said, begins three hundred thousand years after the Big Bang. Let's now go back further, to the origin of the cosmos itself, to the fundamental constants and the 'fine-tuning' argument – all that twiddling to get the knobs in just the right positions. Let's look again at the problem. Beginning with an interesting idea called the anthropic principle.

Anthropos is Greek for 'human'. Hence words like 'anthropology'. We humans exist. We know we exist because here we are, thinking about our own existence. So, the universe we inhabit has to be the kind of universe that is capable of giving rise to us. And the planet we live on has to have the right conditions to give rise to us. It's no accident that we are surrounded by green plants. Any planet lacking green plants (or their equivalent) could not give rise to beings capable of thinking about their own existence. We need green plants as our ultimate food source. It's no accident that we see stars in our sky. A universe without stars would be a universe without any chemical elements heavier than hydrogen and helium. And a universe with only hydrogen and helium would not be rich enough in chemicals to generate the evolution of life. The anthropic principle is almost too obvious to need stating. But it's still important.

Life as we know it needs liquid water. Water exists as a liquid in only a narrow band of temperatures. Too cold and it's solid ice. Too hot and it's gassy vapour. Our planet happens to be just the right distance from our sun, so water can be liquid. Most planets in the universe are either too far from their star (like Pluto – and yes, I know

Pluto isn't classified as a planet any more, but the point remains) or too close to it (like Mercury). Every star has a 'Goldilocks Zone' (not too hot and not too cold but, like Baby Bear's porridge, 'just right'). Earth is in the sun's Goldilocks Zone. Mercury and Pluto, in their two opposite ways, are not. But of course, says the anthropic principle, Earth has to be in the Goldilocks Zone because we exist. And we couldn't exist unless our planet was in the Goldilocks Zone.

Now, what goes for planets would also go for universes. As I've already mentioned, physicists have good reason to suspect that our universe is one of many universes in a 'multiverse'. The multiverse follows – at least according to some interpretations – from the theory called 'inflation', which is accepted by most cosmologists today although it is more 'You cannot be serious' than anything else in science. And there's no reason to suppose that all the billions of universes in the multiverse have the same laws and fundamental constants. The tuning of G, the gravitational constant, could be all over the dial in different universes. It could be that only a small number of universes have their G tuned to the 'sweet spot'. Only a minority of universes are 'Goldilocks universes', whose laws and constants happen to be 'just right' for the eventual evolution of life. And of course (here's the anthropic principle again), we have to be in one of that minority of universes. Our very existence determines that our universe has to be a Goldilocks universe. One friendly Goldilocks universe among possibly billions of unfriendly parallel universes.

You cannot be serious!

It's too early to follow up with *But it's true*. Physicists need to do more work on the problem. What we can say is that it's looking promising. What's more – and this is the main point of my final chapter – the bold step into the frightening void of what seems improbable has turned out right so often in the history of science. I think we should take our courage in both hands, grow up and give up on all gods. Don't you?

Picture acknowledgements

Plate 1: Anders Ryman/Getty. **Plate 2:** Denis-Huot/ naturepl.com. **Plate 3:** Svoboda Pavel/Shutterstock. **Plates 4–6:** Roger T. Hanlon. **Plate 7:** Ron Offermans/Buiten- beeld/Minden/Getty. **Plate 8:** *all* Alex Hyde. **Plate 9:** *top left* Martin Fowler/Shutterstock; *top right* Monontour/ Shutterstock; *middle left* Lisa Mar/Shutterstock; *middle right* Kawongwarin/Shutterstock; *bottom left* Simon Bratt/ Shutterstock; *bottom right* Aprilflower7/Shutterstock. **Plate 10:** *top left* Fiona Stewart; *top right* unknown. **Plate 11** David Tipling/naturepl.com. **Plate 12:** Jill Fantauzza. **Plate 13:** Illustris Simulation/illustris-project.org.

Page 191: Photo and specimen from Carles Millan.

Index

Numbers in **bold** refer to pages with illustrations.

Matthew, gospel of *cont.*
    miracles at moment of
    crucifixion, 38; account
    of virgin birth, 29–30;
    authorship, 21; date
    of, 22, 27; emphasis on
    fulfilment of prophecies,
    29–32, 33–4; official
    canon, 26–7, 33; sayings
    of Jesus, 116, 117–18, 120;
    Sermon on the Mount,
    85; story of fig tree,
    118; story of Gadarene
    Swine, 120–1
Matthew, Patrick, 267
Meir, Golda, 103
memes, 239
Mercury, planet, 277
Messiah, 10, 19–20, 30–1, 60
Micah, 30, 32
miracles: conjurors, 40–2;
    Fatima, 42–5; Hume on,
    39–40, 42, 44–6; Jesus's,
    25, 38–40, 64, 118, 119,
    120–1; sun standing
    still, 46
missionaries, 101
Mohammed, Prophet, 70, 99,
    115, 121
monotheism, 6–7, 8, 48, 78, 229
moral: philosophy, 132–3, 138,
    140; rules, 112, 138–9;
    values, 125–6, 128–9,
    131–2, 140

morality, 96, 126, 140–1, 239
Mormon, Book of, 61–3, 64
Mormonism, 61–4, 70
Moses: authorship of Torah,
    50, 53; Exodus from
    Egypt, 50–2, 82; God
    speaking to, 78, 81,
    107; story of the golden
    calf, 78–9, 105; Ten
    Commandments, 79,
    105, 109
Multiverse Theory, 262,
    270, 277
Muslims, *see* Islam
mutation, 176–80, 222–3
myths: belief in, 69–70;
    development of, 57–60;
    origins of, 56–7, 68

natural selection: babies,
    208–9; compared with
    artificial selection,
    181–2; Darwin's work,
    177, 265–6, 268; DNA,
    209–10, 223; evolution
    by, 154, 169; grazing
    animals, 184; potato
    plant, 183; survival, 223;
    tendency to believe
    parents, 236, 237;
    tendency to niceness,
    239–41, 244, 246;
    tendency to notice
    patterns, 235

ABOUT THE AUTHOR

RICHARD DAWKINS is one of our greatest explainers, thinkers and writers and has been acknowledged as such since publication of his world-altering first book, *The Selfish Gene* (1976), which was voted the Royal Society's 'Most Inspiring Science Book of All Time'. He is also the author of the bestsellers *The Blind Watchmaker, Climbing Mount Improbable, The Ancestor's Tale, The God Delusion* and *The Magic of Reality,* two anthologies, *A Devil's Chaplain* and *Science in the Soul,* and two volumes of autobiography, *An Appetite for Wonder* and *Brief Candle in the Dark.* He is a Fellow of New College, Oxford, and of both the Royal Society and the Royal Society of Literature. In 2013 Dawkins was voted the world's top thinker in *Prospect* magazine's poll of 10,000 readers from over 100 countries.

richarddawkins.net